Data Fitting in the
Chemical Sciences

Data Fitting in the Chemical Sciences

By the Method of Least Squares

PETER GANS
School of Chemistry, The University of Leeds

JOHN WILEY & SONS
Chichester · New York · Brisbane · Toronto · Singapore

Copyright © 1992 by John Wiley & Sons Ltd
Baffins Lane, Chichester
West Sussex PO19 1UD, England

Other Wiley Editorial Offices

John Wiley & Sons, Inc., 605 Third Avenue,
New York, NY 10158-0012, USA

Jacaranda Wiley Ltd, G.P.O. Box 859, Brisbane,
Queensland 4001, Australia

John Wiley & Sons (Canada) Ltd, 22 Worcester Road,
Rexdale, Ontario M9W 1L1, Canada

John Wiley & Sons (SEA) Pte Ltd, 37 Jalan Pemimpin #05-04,
Block B, Union Industrial Building, Singapore 2057

Library of Congress Cataloging-in-Publication Data

Gans, Peter.
  Data fitting in the chemical sciences by the method of least
squares / Peter Gans.
      p.   cm
  Includes bibliographical references and index.
  ISBN 0 471 93412 7
  1. Least squares. 2. Curve fitting. I. Title.
QA275.G25   1992
  511'. 42—dc20   91-39944
                                                   CIP

British Library Cataloguing in Publication Data

A catalogue record for this book is available from the British Library

ISBN 0 471 93412 7

Typeset in Times 10/12pt by Mathematical Composition Setters Ltd, Salisbury, Wiltshire
Printed and bound in Great Britain by Biddles Ltd, Guildford, Surrey

Contents

Contents

Preface

As scientists, we all have to use black boxes. A black box is a device into which we put a sample in order that, more or less mysteriously, we can measure a property of the sample. If we do not understand what is happening in the black box we cannot claim to understand fully what the measurement means. Nowadays many instruments, which may or may not be black boxes, come complete with another black box—a computer. During the 1980s micro-computers became commonplace in the laboratory environment, both inside instruments and external to them. The micros typically both control the instruments and offer a variety of options to process the measurements. This book is about some of the things that go on in the black boxes that process experimental data.

One of the most common processes applied to data is that of least-squares fitting, a powerful procedure which is used to reduce the errors in the experimental measurements. The process is introduced in Chapter 1 and the nature of experimental errors is first discussed in Chapter 2. The general theory is given in Chapters 3, 4 and 5, and this is followed by a critical examination of the validity of the results, including an introduction to methods of statistical testing.

In the second half of the book we look at ways in which the theory is applied to typical data-fitting problems. Fitting with polynomials is examined in Chapter 7, which includes not only fitting entire data sets with a straight line, a parabola, etc., but also the techniques of smoothing and differentiation which involve the fitting of subsets of the data with such functions. Functions covered in Chapter 8 include spline functions, single and multiple exponentials, Gaussian, Lorentzian and related functions, trigonometric functions and surface functions.

Chapter 9 gives a summary of Fourier transform techniques insofar as they impinge on data fitting. This chapter has been included because of the importance of Fourier transform methods in present-day systems of data collection.

Chapter 10 gives the history of a particularly complex system that I have been associated with for many years. Although this is a somewhat specialized application, it does draw together many of the themes presented earlier and gives substance to some of the apparently arcane ideas presented in the theoretical section.

Although I have approached the subject largely from a mathematical point of view, it is not necessary to know much mathematics to understand the reasons for doing what we do. The emphasis is more on why we use certain procedures, and how we may interpret the results. The mathematics is little more than a concise and precise language which is used to explain experimental methodology. In line with this approach I have not talked about any specific computer or programming language—the methodology is not dependent on the hardware or on the software. It is my hope that this approach will help people either to evaluate critically packages that they use, or to specify a computer system to be used on their own data.

The central theme concerns what is going on in the black box. I have tried to summarize my experience of various things that can go wrong, both before and after processing the data, and of the many pitfalls awaiting the unwary when it comes to interpreting the results. My aim is to help people to *use* data rather than to *abuse* it.

I wish particularly to thank: Bernard Gill, with whom techniques have been developed for the analysis of infrared and Raman spectra obtained during our research on solution chemistry; Leslie Pettit for his enthusiastic use, on potentiometric titration data, of MINIQUAD and SUPERQUAD during their development stages; Terry Whitehead for the experimental data in Section 6.8; David Bower for reading Chapter 9. I am also greatly indebted to the many research students who have shared their data-fitting problems with me over the last 20 or so years and who have helped me to understand some of the very real problems that have had to be solved.

All the diagrams in this book were produced by computer programs written for my favourite micro, an Archimedes 440, in BBC BASIC V. I thank Mark Thornton-Pett for the use of his laser printer to print the diagrams. Printouts of some of the programs can be obtained on request. All the 'recipes' have been checked by writing an appropriate program, including a multipurpose least-squares program. However, any errors that remain in the text are to be attributed to me, not to the Archimedes.

1

Introduction

It is commonplace, when you have a number of measurements of a quantity, to take the average as a better estimate of that quantity than any of the individual measured values. Have you ever wondered what the justification is for this procedure and do you know what are its limitations? The question we are asking is, when we have made repeated measurements how can we derive the best estimate of the things that determine the magnitude of the measured values? There are many issues involved in this question, not the least of which is what we mean by the term 'best'. The purpose of this book is to explore all these issues and to develop procedures that can be applied to even the most complicated situations.

To introduce what is involved let us examine an example drawn from everyday life. I set out to answer the following question: do penny coins lose mass after being in circulation for a long time? To this end I collected 200 one penny coins from the bank, sorted them by year of issue, and weighed them on an analytical balance which reads to 0.1 mg. This is stage 1 of the process—data collection. Two points should be noted. Firstly, the coins collected are a small *sample* of all the coins that were minted. We must assume at this stage that the sample is representative of all the coins minted. Secondly, the balance was calibrated against a set of masses, certified by the National Physical Laboratory for use as secondary standards.

New coins do not all have the same mass. Indeed the Royal Mint specifies the mass of 1p coins as 3.564 ± 0.075 g. (The full significance of ± will become apparent in due course).

Let us now look at a set of masses of new coins as shown in the first column of Table 1.1. The first thing to notice is that the masses vary in the second decimal place. The fact that the weighings are given to four decimal places shows that we are measuring the mass to a level of precision that clearly shows up variations from one coin to another. Also, we can see that the error in the

Table 1.1. Masses of new coins

Coin no.	Mass/g	Residual/g	Residual^2/g^2
1	3.5485	−0.0107	0.000 114 63
2	3.5518	−0.0074	0.000 054 86
3	3.5677	0.0085	0.000 072 14
4	3.5763	0.0171	0.000 292 19
5	3.5922	0.0330	0.001 088 57
6	3.5472	−0.0120	0.000 144 15
7	3.6026	0.0434	0.001 883 00
8	3.5918	0.0326	0.001 062 34
9	3.5858	0.0266	0.000 707 22
10	3.5536	−0.0056	0.000 031 43
11	3.5974	0.0382	0.001 458 75
12	3.5446	−0.0146	0.000 213 35
13	3.5532	−0.0060	0.000 036 08
14	3.5096	−0.0496	0.002 460 80
15	3.5650	0.0058	0.000 033 57
16	3.5350	−0.0242	0.000 585 95
17	3.5655	0.0063	0.000 039 61
18	3.5345	−0.0247	0.000 610 41
19	3.5853	0.0261	0.000 680 87
20	3.5945	0.0353	0.001 245 63
21	3.5206	−0.0386	0.001 490 46
22	3.5937	0.0345	0.001 189 80
23	3.5451	−0.0141	0.000 198 99
24	3.6169	0.0577	0.003 328 55
25	3.5251	−0.0341	0.001 163 25
26	3.5153	−0.0439	0.001 927 78
27	3.5276	−0.0316	0.000 998 97
28	3.5486	−0.0106	0.000 112 50
29	3.5656	0.0064	0.000 040 88
30	3.5385	−0.0207	0.000 428 76
31	3.5363	−0.0229	0.000 524 71
Sum	110.3354	0	0.024 220 18
Average	3.5592	0	0.000 807 34
Standard deviation			0.028 413 72

mass measurements is small compared to the variation in coin mass. This is not the normal situation that we face in the physical sciences, where measurement errors are usually the most important. However for the purpose of our illustration this does not matter; the measured quantity is subject to variation and it is only the source of this variation that is unusual.

From the figures given we calculate the *average* as 3.5592 g. This value looks close to what we might expect, 3.564 g. We can be more precise about how close it is if we can ascribe an *error* to this calculated value. The best estimate is probably the *standard deviation*, which is 0.0284 g. We could then say that the mass of a single coin as derived from our weighings is 3.56 ± 0.03 g. Note especially that in the final report we quote the error estimate to one significant figure. This is because the error estimate is itself subject to an error which is probably more than 10% of its value (Section 5.3). Having quoted the error to one significant figure, we must quote the magnitude to no more significant figures than are warranted by the error. This is because the digits after 3.56 are meaningless in the sense that they are subject to complete uncertainty.

We can find out more about the variability of coin mass by looking at the differences between the observed values and the average. These are shown in the second column of Table 1.1, the columns of *residuals*. The residuals are the differences between the observed mass and the calculated mass. Since in this case the calculated quantity, the average, is the same for all observations, the residuals are equal to the differences between the observed values and the average. There are 17 coins that weigh less than the average and 14 that weigh more than the average. This indicates that the mass is subject to *random* variation. The third column shows the values of the residuals squared and the average[*] of the squared residuals. The latter is the *variance* of the measurements and its square root is the *standard deviation*, which was quoted above.

We could describe the variation in coin mass as *experimental error*. The only way in which we could determine the errors would be if we knew the true value for the average of the whole population of coins. It so happens in this instance that we do know this quantity, but in general we do not know it. We must make the fundamental distinction between the errors and residuals. Without knowledge of the true value of a quantity we can only work with residuals.

Let us begin to justify the use of the average and standard deviation. We first define a *model* of the system:

$$\text{Observed mass} = \text{calculated mass} + \text{error}$$

Implicit in this model is the idea that the calculated mass is a constant:

$$\text{Calculated mass} = m_0$$

We then found the estimate of the *parameter*, m_0, that minimizes the variance, or sum of squared residuals. If we denote the observed masses as m_i, meaning the mass of the ith coin, we have minimized the sum S with respect to m_0:

$$S = \sum_{i=1,31} (m_i - m_0)^2$$

The summation is carried out over all 31 observations. The minimum is specified by setting the derivative $\mathrm{d}S/\mathrm{d}m_0$ equal to zero

---

[*] Although there are 31 observations, the sample variance is the sum of squared residuals divided by 30.

(the summation range $i = 1, 31$ is implied):

$$dS/dm_0 = \Sigma - 2(m_i - m_0) = 0$$

$$62m_0 = 2\Sigma m_i$$

$$m_0 = \Sigma m_i / 31$$

This proves that the average is the estimator that minimizes the variance, S. We may also prove that the sum of the residuals in this case is zero (the general proof is in Section 6.4):

$$\Sigma (m_i - m_0) = \Sigma m_i - 31m_0 = 0$$

The proof that the standard deviation is a good estimate of the error on the parameter is given in Appendix 6.

By minimizing the sum of squared residuals we have obtained the so-called least-squares estimate of the mass of a coin, which is always a minimum variance estimate, but we have still not addressed the more fundamental question as to why we have minimized this function. To answer this fully we would have to delve deep into the realm of statistics, so for the moment we shall say that, subject to the condition that the experimental errors belong to a *normal distribution*, the minimum variance result is also the *maximum likelihood* result, which is to say the result with the greatest probability of being near to the correct result; a fuller explanation of statistical terms is given in Appendix 1. It is often the case that the actual experimental errors are drawn from an approximately normally distributed population, and so there is some justification for using the least-squares method. However, the least-squares method is so powerful that it is often used where there is no knowledge of the statistics of the experimental errors; it is blindly assumed that it will give a reasonable result! Fortunately there are not many cases where the least-squares method gives a bad result.

According to the theory given in Section 6.4.1, if the experimental errors belong to a normal distribution certain scaled residuals should also belong to a normal distribution. In this case the scaling factors are constant, so our residuals should belong to a normal distribution. To see whether they do so or not we can draw a histogram to represent the frequency of occurrence. The residuals were grouped into 14 groups according to magnitude. The number in each group or frequency of occurrence is plotted as a histogram in Figure 1.1. Each vertical rectangle has the width of the group, or range of residual values, and a height equal to the frequency of occurrence. Also plotted is the curve for a normal distribution, having the same average and standard deviation as the residuals themselves. It can be seen that the residuals are approximately normally distributed. The agreement is not spectacularly good, but the number of data points is probably too small to expect any better.

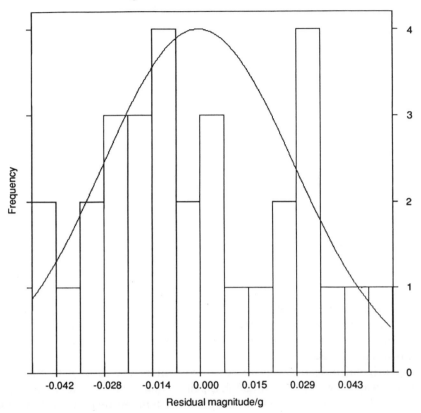

Histogram of residuals of 1989 coin masses

Figure 1.1. Histogram of residuals of 1989 coin masses. The curve represents the normal distribution of mean zero and same standard deviation as the residuals

The best that we can say is that the experimental errors are probably not far from being normally distributed.

To examine the effect of age on coin mass I used the data given in Table 1.2. The data goes back to 1971, the first year in which the new 1p coins were issued, and there are gaps where I had been unable to collect a reasonably large sample of coins from that year. The average of these 124 measurements is 3.5605 g and the standard deviation is 0.0413 g. A histogram of the residuals is shown in Figure 1.2. Also plotted on the figure is the curve corresponding to a normal distribution with the same mean and standard deviation as the coin mass residuals. The maximum value of the normal distribution was arbitrarily chosen as 12. These residuals seem approximately normally distributed, but the actual distribution is somewhat skewed, i.e. not symmetrical about zero. I then calculated the averages and standard deviations for the coin

Table 1.2. Coin masses

Year				Mass/g				
1971	3.5316	3.5470	3.5227	3.5560	3.5345	3.5732	3.5618	3.5905	3.5732
1974	3.5137	3.4633	3.6159	3.5257	3.5535	3.5951	3.5899	3.5875	3.5579
1976	3.5085	3.6633	3.5109	3.4570	3.5057	3.6554	3.5965	3.6010	3.5397
1978	3.5583	3.5744	3.5432	3.5486	3.5510	3.6030			
1979	3.5574	3.5682	3.5840	3.5726	3.5857	3.5908	3.5380	3.5831	3.4590
1980	3.5319	3.5474	3.5462	3.5987	3.5807	3.5210	3.5193	3.5431	3.5986
1981	3.5614	3.5629	3.5384	3.5606	3.4621	3.5940	3.5519	3.6213	3.5794
1983	3.6224	3.5932	3.6019	3.6124	3.5078	3.5933	3.6462	3.5844	3.5549
	3.5676	3.6177	3.5464						
1984	3.6391	3.6029	3.5424	3.4491	3.5865	3.5424	3.5582	3.5890	3.5795
1985	3.5312	3.5619	3.5585	3.6358	3.6040	3.5771	3.5731		
1986	3.4904	3.6031	3.5698	3.4749	3.5540	3.5384	3.5489	3.6500	3.5546
1987	3.5090	3.5524	3.5552	3.5354	3.6059	3.5313	3.5475	3.5131	3.6123
1988	3.4938	3.4608	3.5833	3.5618	3.5605	3.5657	3.5057	3.5831	3.6182
1989	3.5392	3.5663	3.5281	3.5492	3.5161	3.5255	3.6174	3.5456	3.5944

mass from each year. The results of this calculation are shown in Table 1.3. The average mass for each year is plotted in Figure 1.3, together with error bars whose width is one standard deviation. Looking at the figure, the first thing that strikes one is the relatively large uncertainty in the average mass for each year. This will have to be taken into account when we are looking for trends in coin mass.

Our original objective was to see if coin mass decreases with time. The simplest way in which this might happen is that there should be a regular decrease. This is to be our new model of the system. At this stage it is a model without a theoretical basis; it is an empirical model, chosen for the merit of its simplicity.* We must now express the model in mathematical terms: the mass of a coin is equal to its initial mass plus an amount calculated by multiplying the rate of mass change per year by the number of years:

$$\text{Calculated mass}_i = m_0 + \text{rate} \times t_i$$

Here we designate the initial mass as m_0 and the number of years as the time t_i. The initial mass and the rate of change are both unknown quantities, so we have two parameters. The situation is more complicated than before when m_0 was the only unknown quantity. Not only do we have two parameters, but there is a new variable t which is an *independent variable*. We are assuming that the age of a coin is known exactly or, put another way, the variable t is free from error. We can use the method of least squares to determine both m_0

---

*The two most obvious mechanisms for mass change are corrosion and wear. A more sophisticated model would take these mechanisms into account.

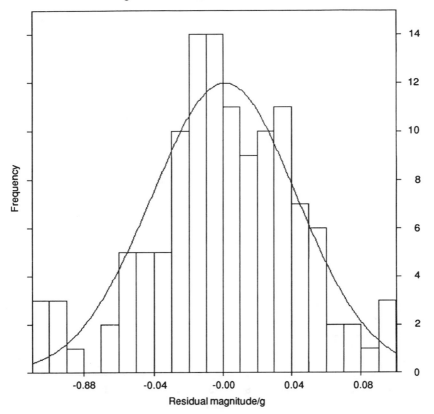

Histogram of residuals of coin masses 1971–1989

Figure 1.2. Histogram of residuals of coin masses 1971–1989. The curve represents the normal distribution of mean zero and same standard deviation as the residuals

Table 1.3. Average coin masses and errors

Age(y)	Average mass/g	Estimated standard deviation/g
0	3.5535	0.0317
1	3.5481	0.0476
2	3.5513	0.0344
3	3.5538	0.0499
4	3.5774	0.0313
5	3.5655	0.0502
6	3.5873	0.0365
8	3.5591	0.0414
9	3.5541	0.0293
10	3.5599	0.0389
11	3.5631	0.0204
13	3.5594	0.0676

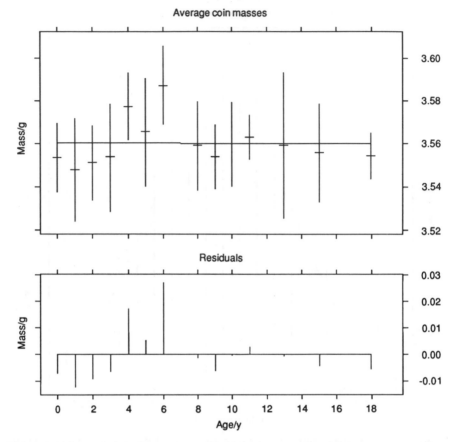

Figure 1.3. Best straight line fit to average coin masses 1971–1989. Upper box: experimental values with error bars of 1 standard deviation and best-fit straight line. Lower box: residuals

and rate. How this is done is explained in the following chapters. For the moment we want to look at the results, obtained by the methods of Section 7.6 and see how they can be interpreted. Taking the weights equal to the reciprocal of the square of the estimated standard deviations, which is a way of making allowance for the fact that the observed average mass for each year is subject to some uncertainty, the best straight line through the average coin masses, also shown in Figure 1.3, had the following values for the parameters:[*]

$$m_0 = \quad 3.5625 \pm 0.005159 \text{ g}$$
$$\text{Rate} = -0.00024 \pm 0.000486 \text{ g y}^{-1}$$

---

[*] The results are shown here to more significant figures than we would give in the final report in order that we may subsequently calculate the statistics correctly.

It is obvious that the best straight line is almost flat i.e. that the rate of mass decrease is near to zero. There is a small decrease, but is it significant? The answer to that question takes us into the realm of statistics.

The statistical analysis is based on the assumption that the experimental errors belong to a normal distribution. Since the histogram shows that this assumption is of doubtful validity we must be very cautious about the interpretation of the statistics. Nevertheless, the tests will be performed as a way of introducing statistical testing.

The first test is a qualitative one. The standard deviation of the rate parameter (0.000486 g y^{-1}) is larger than the absolute value (0.00024 g y^{-1}), so we must conclude that the rate is somewhere near to zero. To make this quantitative we may perform a t-test on the ratio of the value to the standard deviation (Section 6.5). This ratio is 0.49. Fourteen points were used to calculate the two parameters, so there are 12 degrees of freedom. We may test the *null hypothesis:*

$$H_0: \qquad Rate = 0$$

Since $t_{0.6} = 0.26$ and $t_{0.7} = 0.54$, we conclude that there is roughly a 30% chance that we will be making a mistake if we reject the hypothesis (see Section 6.6). This chance is too high for comfort, so it is best to accept the hypothesis. In other words, the experimental data do not provide substantive evidence that the mass of coins varies with time. As confirmation of this the value of m_0 when the rate is taken as zero, i.e. the value obtained with the first model (3.5605 g), is not significantly different from the value obtained with the second model (3.5625 g).

The residuals for the two-parameter model, also shown in Figure 1.3, do not seem to be distributed randomly, so it is legitimate to ask whether the straight line is a satisfactory model for these data. To answer this question we compare the weighted sum of squares (1.1791) with $\chi^2/12$ for 12 degrees of freedom. The latter has the same value at the 29.1% significance level. This is the probability that we would be wrong, in rejecting the model, in deducing that the data did not belong on a straight line.

The large residual at 6 years appears to be dominant. What would happen if we were to ignore this datum point? Nominating this point as an *outlier* (Section 5.4), we could recalculate the parameters of the two models:

$$m_0 = 3.5595 \pm 0.004\,024 \text{ g}$$
$$Rate = -0.000\,09 \pm 0.000\,371 \text{ g y}^{-1}$$

The results are not significantly different, so the data for 6-year-old coins need not be dismissed. Moreover, sampling theory can be used to show that there is a very high probability that the deviation of this point from the straight line is due to sampling error; this could be checked by examining a larger sample of coins.

Finally, we should note that the errors on the average masses are rather large in relation to the amount of variation in the averages themselves. The best straight line passes through all the error bars. From that point of view we should regard the model as very satisfactory.

Let us now try to take an overview of the least-squares method of analysis of data. The process of data analysis can be thought of as having a number of distinct aspects. Firstly there is the collection of the data. Then there is the formulation of a model relating the numerical values obtained to one or more parameters and to one or more independent variables. Parameters are the unknown quantities in the model, whereas the independent variables are 'known' quantities. Using the model, one then obtains estimates of the parameters and their associated errors, together with subsidiary information such as the differences between the values calculated with the model and the values obtained by observation. It is good practice to make a graph of the model which shows the experimental data, the calculated data and the residuals. Finally, the model is assessed in terms of such criteria as goodness of fit and relation to underlying mechanisms. Indeed, one objective of the analysis may well be to try to distinguish between various possible mechanisms. After a preliminary analysis the whole process may be repeated. More data may be collected or the method of data collection may be improved. A new model may be postulated and compared to the previous model. Finally, the best model is selected and used to throw light on the underlying mechanisms.

2

Observational Errors

2.1

2.1.1 SYSTEMATIC AND RANDOM ERRORS

All observations are subject to errors of various kinds. One of the main objectives of the least-squares method is to reduce the effect of experimental errors on the calculated values of the parameters, the estimation of which is the true object of the experiments in which the observations were made. We must therefore try to understand the origin and nature of the errors in observations.

Observational errors can be conceptually divided into two categories, systematic and random errors. To emphasize the fact that this classification is conceptual we must be aware that we cannot measure the errors themselves, unless we know the 'true' values of the quantities being measured, and, if we know the 'true' values there is no point in making the measurements. In the world of experimental data, therefore, the distinction between systematic and random errors is not clear-cut.

Systematic errors are those errors which are present in all measurements. For example, in measuring coin mass the balance may not read exactly zero when there is nothing on it. (In the example given in Chapter 1 the balance was zeroed at regular intervals, but the zero drifted by up to 0.5 mg in between.) The zero error will affect all weight measurements and so it is a constant, or systematic, error. On the other hand, the balance may not respond exactly linearly to increases in mass, in which case the error will vary with the mass being measured and there will be a varying systematic error.

The least-squares method is incapable of telling us anything about systematic errors because they are common to all data points. Therefore it is up to the experimentalist to reduce the systematic errors as much as possible, e.g. by calibrating the apparatus against secondary or even primary standards. The calibrations should ideally be performed both before and after the experimental measurements. The experimental readings are then automatically or manually corrected for systematic errors. Unfortunately this process requires the experimentalist to be able to identify all sources of systematic error, and this is not always easy or practicable. For instance, the performance of a balance may be affected by such external influences as temperature, humidity and atmospheric pressure and so forth. It must therefore be accepted that even after taking all reasonable precautions the experimental data will always be subject to some systematic error.

Random errors are inherent in all measurements. They are the errors remaining after all systematic error has been eliminated. In view of the impossibility of completely eliminating systematic error we must accept that the concept of random error is idealistic, but the situation does often arise that random errors are larger than systematic errors so that the latter can, to a first approximation, be ignored. Indeed, a realistic goal for the reduction of systematic error is to make it negligible with respect to random error. This goal was achieved in the coin mass experiment by using a high-precision balance.

2.1.2 SOURCES OF ERROR

Some of the most common sources of error are as follows. This list is not exhaustive as there are almost as many types of error as there are experiments (Currie, 1978).

(a) Blunders. These are mistakes in reading instruments or in copying values from one book to another, such that the value finally recorded does not correspond to the correct measurement. Although with care blunders can be made to occur rarely, one should be aware that blunders cannot be totally eliminated. With digital instruments there may be transmission errors. A common source of such errors is the presence of spikes on the electricity supply which cause transient malfunctions in the electronics.

(b) Sampling error. Any process that requires the taking of a sample in order to establish a property of a bulk material is subject to the error that the bulk material is hardly ever uniform, so that samples vary in composition. Sampling error depends on the number of samples taken (Section 5.3) and can be reduced by taking more samples.

In determining the concentration of trace elements in seawater, early measurements were erroneous because of contamination by the sampling equipment itself. This is another example of sampling error.

(c) Bias. When humans try to read off between scale marks on an instrument they usually favour certain values over others. This introduces a bias into the readings which has been called 'the human equation'. However, digital instruments may also be biased in the last significant digit, e.g. they may round down to the next lowest number rather than round to the nearest number.

(d) Calibration error. All measurement systems depend ultimately on reference to internationally agreed primary standards. Usually they are calibrated against secondary standards which in turn are calibrated against the primary standards. Standards are usually designed so as to make calibration error negligible, but as the precision of experimental measurements improves so calibration error will assume more importance. For instance, in many chemical measurements the calibration standard is a sample of something that is easy to purify. The calibration error will then depend on the actual purity of the sample.

(e) Electronic noise. All electronic circuits are imperfect since electrons are subject to the uncertainty principle of quantum mechanics. Variations in the performance of electronic components introduces errors into the data. An example of electronic noise is the 'snow' that appears on a domestic television screen when the signal is weak or the set is not properly tuned.

2.1.3   ACCURACY AND PRECISION

The presence of errors in experimental data will clearly affect the reliability of any parameters obtained by the least-squares method. We say that systematic errors affect the *accuracy* of the results, while random errors affect the *precision* of the results. This distinction between accuracy and precision is of fundamental importance: it is useless to reduce random errors with uncalibrated or inaccurate equipment. Accurate results can only be achieved by reducing both systematic and random errors.

The precision of results depends only on the nature and magnitude of random errors. Clearly the more the magnitude of random errors is reduced the greater will be the precision. One simple rule to follow is that one should always try to reduce the magnitude of the largest error before dealing with the smaller ones. This follows from the law of propagation of errors (see Section 2.2).

2.1.4   ESTIMATION OF RANDOM EXPERIMENTAL ERROR

The most convenient descriptors of random error are the variance, σ^2, which is the square of the actual error, and the standard deviation, σ, which is the positive square root of the variance. These terms are explained in Appendix 1. Note that all estimates of error are by definition positive.

Variance may be estimated experimentally by taking repeated measurements. If we make m measurements of a quantity y the variance of this sample of all the measurements that could be taken—the sample variance—given by

$$s^2 = \frac{\Sigma(y_i - \bar{y})^2}{m - 1}; \qquad \bar{y} = \frac{\Sigma y_i}{m}, \qquad \Sigma \equiv \sum_{i=1,m}$$

is an unbiased estimate of the variance of y (Appendix 1). However, this estimate is itself subject to error (Section 5.3), so it should be treated conservatively.

When repeated measurements are not possible we have to guess the errors. A universal source of random error that can be guessed is the finite number of digits that can be read off an instrument scale—the instrument precision. For example, suppose a digital voltmeter gives a reading in millivolts; there will be a random error of about 0.5 millivolts in each reading. This shows that it is possible to estimate the magnitude of a random error by consideration of the instrument performance, even when the instruments are not supplied with information on specifications and tolerances.

Sometimes an estimate of error is implicit in the way that the measurement is made. For example, if one decides to read a 50 ml burette to the nearest division, one is deciding that the error on the burette reading will be about 0.05 ml, but if one decides to try to read to a tenth of a division the error is about 0.02 ml.

It is important to realize that errors arise in the measurement system as a whole, and not just in one component of the system. To illustrate this, when making precise measurements of mass, the samples should be handled with tongs, as use of the fingers can add small amounts of moisture to the sample; balance doors should be closed to eliminate draughts; dry samples should be stored in a desiccator and so on. The final estimate of error should take into account all potential sources of error, combining them together as indicated in Section 2.2.

Variance relates to the error on a single quantity. There is also the possibility that the error on one measurement will depend on the error on another. To deal with this possibility we introduce the idea of a covariance. From the definition of covariance (Appendix 1) it can be estimated from m repeated measurements:

$$\text{COV}(y_i, y_j) = \frac{\Sigma_{k=1,m}(y_{ik} - \bar{y}_i)(y_{jk} - \bar{y}_j)}{m - 1}$$

Covariance is often rather difficult to measure experimentally, and for this reason it is often ignored. It is a good experimental practice to try to reduce covariances to zero. This means trying to ensure that errors on separate measurements are independent of each other. A common case is one in which measurements are taken at a sequence of different times: if sufficient time is allowed between readings for the measuring system to settle down it will give reliable estimates of the error, but if insufficient time is allowed one reading will depend on the previous reading, the one before that and so forth, the effect getting smaller as the distance from the current reading increases. It should be noted that in many modern instruments the raw measurements may be processed numerically in such a way that the data output has non-zero covariance (see Section 7.4.6).

Covariance is related to the correlation coefficient by

$$(\text{COV})_{ij} = \sigma_i \sigma_j \rho_{ij}$$

A correlation coefficient can only take values between $+1$ and -1. The correlation coefficients ρ_{ii} are 1 since the error of any quantity is completely correlated with itself. Together the variances and covariances make up the so-called variance–covariance matrix whose elements are given by

$$\mathbf{COV} = \begin{bmatrix} \sigma_1^2 & \sigma_1\sigma_2\rho_{12} & \sigma_1\rho_3\rho_{13} & \cdots \\ \sigma_1\sigma_2\rho_{12} & \sigma_2^2 & \sigma_2\sigma_3\rho_{13} & \cdots \\ \sigma_1\sigma_3\rho_{13} & \sigma_2\sigma_3\rho_{23} & \sigma_3^2 & \cdots \\ \cdots & \cdots & \cdots & \cdots \end{bmatrix}$$

When the errors in the observations are independent of each other the variance–covariance matrix takes the simple form $\text{COV}_{ii} = \sigma_1^2$ and $\text{COV}_{ij} = 0$ for $i \neq j$. It is said to be diagonal. If the covariances cannot be reduced to zero by experimental means the matrix of correlation coefficients can be guessed as something of the form

$$\rho = \begin{bmatrix} 1 & 0.5 & 0.25 & \cdots & \cdots \\ 0.5 & 1 & 0.5 & 0.25 & \cdots \\ 0.25 & 0.5 & 1 & 0.5 & \cdots \\ \cdots & \cdots & \cdots & \cdots & \cdots \end{bmatrix}$$

This form of correlation coefficient matrix implies that the correlation coefficients between any pair of successive readings is 0.5 but that the error in each reading also depends on the error of the previous reading. This type of correlation coefficient matrix is used to treat electron diffraction data.

A complete description of experimental error requires the specification of its probability distribution function (Section 6.1). However, estimation of this probability distribution function is extremely difficult. The only case known with some certainty involves radioactive decay, where counting errors follow a Poisson distribution. By default we usually assume that our errors follow a normal distribution, and in many cases this is not a bad assumption. However, unless it can be confirmed by experiment that the probability distribution function is normal, it should be recognized that erroneous conclusions could be drawn when making certain statistical tests (see Chapter 6).

It cannot be stressed too strongly that the estimation of experimental error is just as important as the taking of measurements. Not only do the errors specify the precision of the measurements but they also determine the precision of all the quantities derived from the measurements. Experimental error defines the scale on which the significance of results may be assessed. In particular, the variance–covariance matrix plays an essential role in the least-squares process; when it has been estimated correctly it is possible to apply statistical tests (Section 6.6) that could not otherwise be applied.

2.2

2.2.1 PROPAGATION OF ERRORS IN LINEAR COMBINATIONS

Given that the observations are subject to error, the question arises, how does one calculate the errors on quantities derived from combinations of the observations? We begin with the case of linear combinations and deal with non-linear combinations in the following section. The procedure can be applied to the determination of the error on any combination of quantities subject to error.

Generally, each observation y_i has an associated variance M_{ii}, and each pair of observations y_i and y_j has an associated covariance M_{ij}. The elements M_{ij} constitute the variance–covariance matrix $\mathbf{M}$.[*] Suppose that a set of variables, $\mathbf{p}$, is a linear combination of the observations with a set of transformation constants T_{ji}:

$$p_j = \sum_{i=1,m} T_{ji} y_i; \qquad \mathbf{p} = \mathbf{Ty}$$

Then, it can be shown (Appendix 6) that the variance–covariance matrix on $\mathbf{p}$ is simply given by

$$P_{ij} = \sum_{h=1,m} \sum_{k=1,m} T_{ih} M_{hk} T_{jk}; \qquad \mathbf{P} = \mathbf{TMT}^{\mathrm{T}}$$

P_{jj} is the variance on p_j and P_{ij} is the covariance between p_i and p_j.

Suppose that we wish to calculate the error on some linear combination of two variables x and y:

$$z = ax + by$$

where a and b are constants. Applying the general formula, it is easy to write down the result:

$$\sigma_z^2 = a^2 \sigma_x^2 + b^2 \sigma_y^2 + 2ab(\mathrm{COV})_{xy}$$

This formula can be generalized. For example, given that 1p coins weigh 3.564 ± 0.075 g, what is the calculated error on the mass of 100 coins? The answer is obtained as follows. Let z be the mass of 100 coins. Then

$$z = \sum_{i=1,100} m_i$$

$$\sigma_z^2 = \Sigma 1^2 \sigma_m^2 = 100 \sigma_m^2$$

---

[*] A summary of matrix notation is given in Appendix 2. In this section expressions are given in both matrix and algebraic forms; later in the book only matrix notation will be used.

$$\sigma_z = 10\sigma = 0.75 \text{ g}$$

One may see from this that banks are unlikely to miss the absence of a penny in a bag of 100 coins if the bag is weighed to a precision of 0.1 g.

2.2.2   PROPAGATION OF ERRORS IN NON-LINEAR COMBINATIONS

When we wish to calculate the error on a non-linear combination of quantities subject to error we have to make the assumption that the errors are relatively small. The reason for this is that we have to linearize the function by means of a truncated Taylor series;

$$p_j = f(x) = p_j^0 + \Sigma x_i \, \partial p_j / \partial x_i + \cdots$$
$$\approx p_j^0 + \Sigma T_{ji} x_i$$

and the truncation will only be valid when the errors are small. In the physical sciences this restriction is not severe as we can usually arrange the measuring system to ensure reasonable precision. Indeed, it can be argued that there is little point in trying to obtain a physical quantity unless it can be determined with a good degree of precision.

In practice we use the same formalism as in the case of a linear combination of parameters, with the difference that the transformation coefficient T_{ji} is replaced by the partial derivative $\partial p_j / \partial x_i$; p_j^0 is a fixed quantity and is not subject to error. It follows that

$$P_{ij} = \sum_{h=1,m} \sum_{k=1,m} \frac{\partial p_i}{\partial x_h} \frac{\partial p_j}{\partial x_k} M_{hk}$$

In this way we obtain the variance–covariance matrix of non-linear combinations. It is useful to have expressions for combinations of two quantities:

$$z = f(x, y)$$

From the general formula we obtain

$$\sigma_z^2 = \left(\frac{\partial z}{\partial x}\right)^2 \sigma_x^2 + \left(\frac{\partial z}{\partial y}\right)^2 \sigma_y^2 + 2 \frac{\partial z}{\partial x} \frac{\partial z}{\partial y} (\text{COV})_{xy}$$

This formula is valid for both linear and non-linear combinations, and is used to derive the standard formulae in the following section.

2.2.3   SOME STANDARD ERROR PROPAGATION FORMULAE

In the following formulae x and y are subject to error, but a and b are not:

(a) $z = ax + by$　　　　　　$\sigma_z^2 = a^2 \sigma_x^2 + b^2 \sigma_y^2 + 2ab(\text{COV})_{xy}$
(b) $z = ax - by$　　　　　　$\sigma_z^2 = a^2 \sigma_x^2 + b^2 \sigma_y^2 - 2ab(\text{COV})_{xy}$

(c) $z = axy$ $\qquad$ $(\sigma_z/z)^2 = (\sigma_x/x)^2 + (\sigma_y/y)^2 + 2(COV)_{xy}/(xy)$

(d) $z = ax/y$ $\qquad$ $(\sigma_z/z)^2 = (\sigma_x/x)^2 + (\sigma_y/y)^2 - 2(COV)_{xy}/(xy)$

(e) $z = ax^b$ $\qquad$ $\sigma_z/z = b(\sigma_x/x)$

(f) $z = ax^{-b}$ $\qquad$ $\sigma_z/z = b(\sigma_x/x)$

(g) $z = a\,e^{bx}$ $\qquad$ $\sigma_z/z = b\sigma_x$

(h) $z = a\,e^{-bx}$ $\qquad$ $\sigma_z/z = b\sigma_x$

(i) $z = a^{bx}$ $\qquad$ $\sigma_z/z = b\,\log_e a\,\sigma_x$

(j) $z = a^{-bx}$ $\qquad$ $\sigma_z/z = b\,\log_e a\,\sigma_x$

(k) $z = a\,\log_e(bx)$ $\qquad$ $\sigma_z = a(\sigma_x/x)$

(l) $z = a\,\log_e(-bx)$ $\qquad$ $\sigma_z = a(\sigma_x/x)$

The last eight examples are not so much combinations of parameters as non-linear transformations; the general equation can readily be applied to other examples.

Let us now look at the implications of these results. With addition and subtraction the variances are additive and the effect of covariance will be to increase or decrease the variance of z according to the signs of a, b and COV. With multiplication and division the *relative* variances are additive, so in this case it is not the absolute magnitude of the errors that counts, but the ratio of the error to the parameter. We see that the error on the logarithm of a number is equal to the relative error on that number and so on.

2.2.4 ERRORS AND EXPERIMENTAL DESIGN

The laws of error propagation give us a way of assessing the overall error resulting from contributions from different sources. There is a lesson to learn here concerning the design of experiments. The variances (absolute or relative) are additive. Therefore a single large error tends to swamp the smaller errors, because error is squared in variance. The best way to reduce overall errors is to reduce the largest one; indeed, it is somewhat pointless to reduce small errors whilst leaving the large unchanged, as the resultant error will hardly be affected.

To illustrate this point let us consider an analytical determination of the hardness of water by titration of a 25 ml sample with a standardized solution of EDTA. The amount of EDTA required is termed the titre, which we may also suppose is 25 ml. The hardness may be expressed as

$$\text{Hardness} = \frac{\text{titre} \times M}{\text{volume taken}}; \qquad h = \frac{Mt}{v}$$

M is the known concentration of EDTA which is assumed to have negligible error. The error propagation formulae can be applied to deduce the error on the hardness result:

$$(\sigma_h/h)^2 = (\sigma_t/t)^2 + (\sigma_v/v)^2$$

We can take the covariance term as zero as the measurements of titre and volume taken are completely independent. Let us now assume that σ_v is 0.01 ml and that the titre error is 0.1 ml (a common error value for untrained chemists such as first year undergraduates).

	Titre/ml	Volume taken/ml
Value	25.0	25.00
σ	0.1	0.01
σ/p	0.004	0.0004
$(\sigma/p)^2$	0.000 016	0.000 000 16

$$(\sigma_t/t)^2 + (\sigma_v/v)^2 = 0.000\ 016\ 16$$
$$\sigma_h/h = 0.004\ 02$$

This result shows that the error in the calculated hardness is entirely determined by the error in the titre. If, on the other hand, the titre were estimated to 0.01 ml the error would depend equally on the errors in titre and volume taken; in this case the relative error would be about 0.0006.

A good rule of thumb is that the number of significant figures in an answer should not be greater than the smallest number of significant figures in any of the quantities combined to produce that answer. In the example above the titre was estimated to three significant figures and the volume to four. The hardness should only be quoted to three figures. The justification for this rule is the law of error propagation. Accordingly, if an answer is required to have four significant figures *all* the measurements must be made to that precision, including the 'known' molarity of the EDTA. In this way the error on the least significant digit should be less than 10.

When one wishes to put a number to the *accuracy* of a result, one must combine the estimates of random and systematic errors on that result. Following the logic of the additivity of variance, if one seeks an accuracy comparable to the precision one must seek to reduce the systematic errors to the same level as the random errors. As with random error, the level of systematic error will be principally determined by the largest error, so reducing the largest error will make the greatest contribution to the overall improvement of accuracy.

3

Linear Least Squares

3.1 THE OBJECTIVE FUNCTION

The least-squares method consists of finding those parameters that minimize an *objective* function. It is called objective because it is the object of the method to minimize this function. The objective function depends in some way on the parameters, on a series of experimentally measured numbers, or observations, and on a set of independent variables which are also experimental in origin. The methods used to minimize the objective function belong to the class of methods used to minimize general functions. However, the least-squares objective functions have a particular form which lends itself to particular methods of minimization. The two principal methods will be presented in this chapter.

Suppose that there are m observations. Let us number them $1, 2, 3, ..., m$. Then we allocate them to a vector $\mathbf{y}^{obs}$, the first observation to y_1^{obs}, the second to y_2^{obs}, and so on. We may now say that the m observations are allocated to y_i^{obs} $(i = 1, m)$. Similarly, we may allocate the values of the independent variables to a vector $\mathbf{x}$, x_k $(k = 1, m')$. Notice that the number of independent variables need not be the same as the number of observations, though it very often is. Suppose that there are n parameters. Their values will be obtained from a vector $\mathbf{p}$, p_j $(j = 1, n)$.

We must now define a *model* such that corresponding to each observation, y_i^{obs}, there will be a calculated value y_i^{calc}:

$$y_i^{obs} = y_i^{calc} + e_i; \qquad i = 1, m$$

The terms e_i are the experimental errors. Implicit in this model is the assumption that y_i^{calc} is free from error. The calculated value will depend on the parameters and the independent variables:

$$y_i^{calc} = f(\mathbf{p}, \mathbf{x}_i)$$

x_i denotes a vector of all those values of the independent variable(s) upon which the ith observation depends. Most commonly there is only one independent variable, and we would write x_i for that scalar quantity.

A linear model is one in which the calculated values are linear functions of the parameters, i.e. each parameter is simply multiplied by a function of x, $f(x)$. The general form of a linear model is as follows:

$$y_i^{calc} = \sum_{j=1,n} f_j(x_i)p_j; \qquad i = 1, m$$

Linear models are discussed in this chapter and non-linear models in Chapter 4.

We may now define an objective function. Firstly, we set up a vector of residuals, $\mathbf{r}$:

$$r_i = y_i^{obs} - y_i^{calc}; \qquad i = 1, m$$

The simplest form of objective function is a sum of squared residuals—the term 'least squares' derives from this kind of function:

$$S = \sum_{i=1,m} r_i^2; \qquad S = \mathbf{r}^T\mathbf{r}$$

However, in defining the objective function in this way we are making two important assumptions. The first assumption is clear from the definition: we are assuming that y_i^{obs}, and only y_i^{obs}, are subject to experimental error. It follows that we are assuming that the independent variable is not subject to error. The second assumption is not so obvious. It is that the errors, e_i, are all equal and that they are independent of each other. Minimization for this objective function is discussed in Section 3.2.1.

To be more general, we must accept that each observation is subject to a different error and that the errors for each pair of observations are connected through a covariance term. The variances and covariances are allocated to M_{hk}; $h = 1, m$; $k = 1, m$. In this case we define the objective function as follows:

$$S = \sum_{h=1,m} \sum_{k=1,m} r_h W_{hk} r_k; \qquad S = \mathbf{r}^T\mathbf{W}\mathbf{r}$$

The matrix $\mathbf{W}$ is known as the weight matrix and is the inverse of the variance–covariance matrix $\mathbf{M}$:

$$\mathbf{WM} = \mathbf{MW} = \mathbf{I}; \qquad \sum_{h=1,m} W_{ih}M_{hk} = 1 \quad (i = k); \qquad \sum_{h=1,m} W_{ih}M_{hk} = 0 \quad (i \neq k)$$

It will be seen that this objective function becomes equal to the first in the special case $W_{hh} = 1(h = 1, m)$, $W_{hk} = 0$ $(h \neq k)$. Minimization of this objective function is discussed in Section 3.3.1. A more general case still is one in which we assume that the independent variable may also be subject to error. This is discussed in Section 3.4.1.

Before going into the details of the method it is well to be explicit about the reasons behind what we are doing. We are trying to minimize some function of the residuals, i.e. we are trying to get the best possible agreement between the observations and the calculated values. The function chosen for minimization is a linear combination of residuals. It can be shown that of all the possible linear combinations, that in which the combination coefficients are the weights as defined above gives the lowest value for the objective function. The proof of this assertion in the most general case is embedded in the algebra of Section 3.4; Hamilton (1964) gives the simpler proof applicable when the parameter is the sample average. Minimization of the objective functions discussed here gives the so-called minimum variance estimate of the parameters. The term 'best', as in best fit, can now be understood to imply the lowest possible value of the objective function.

The least-squares method can be applied to any model, regardless of the form of the distribution function of the experimental errors. However, if one wants to apply statistical tests that distribution function must be known, or at least guessed.

There is one case in which the minimum variance estimate of the parameters is also the so-called maximum likelihood estimate. This is the case in which the errors belong to a *normal* distribution. Since experimental errors are often distributed normally, or nearly so, we have in this case an additional justification for using the least-squares method; the values of the parameters obtained will be simultaneously minimum variance and maximum likelihood estimates.

In some circumstances the least-squares criterion can be replaced by another, such as

$$S = |(y_i^{\text{obs}} - y_i^{\text{calc}})|_{\max}$$

This is the so-called minimax criterion, in which the maximum absolute discrepancy between observed and calculated values is minimized. We only discuss the least-squares method in this book; other objective functions can be minimized by general minimization techniques. The correct choice of objective function is crucial to obtaining meaningful results.

3.2

3.2.1   THE UNIT-WEIGHTED NORMAL EQUATIONS METHOD

As we have seen in Section 3.1, the general form of the objective function is

$$S = \sum_{h=1,m} \sum_{k=1,m} r_h W_{hk} r_k$$

where r_k is the kth residual, the difference between the observed and calculated values. When the variance-covariance matrix of the observations is a diagonal, constant matrix, i.e.

$$M_{ii} = \sigma^2, \qquad M_{ij} = 0 \qquad \text{for } i \neq j$$

we can say that it is a unit matrix multiplied by the constant σ^2, known as the variance of an observation of unit weight. Thus, if we take the weight matrix to be a unit matrix, we have

$$W_{ii} = 1, \qquad W_{ij} = 0 \qquad \text{for } i \neq j$$

$$\mathbf{W} = \frac{1}{\sigma^2} \, \mathbf{M}^{-1} = \mathbf{I}$$

and the weights are proportional to the inverse of $\mathbf{M}$. The objective function becomes the sum of squared residuals:

$$S = \sum_{i=1,m} r_i^2$$

The minimum of a function of many variables is specified when the *partial derivatives* of the function with respect to each parameter are zero. (In the case of a single parameter the derivative dS/dp should be zero.) Each of the n partial derivatives is set equal to zero:

$$\frac{\partial S}{\partial p_k} = \sum_{i=1,m} 2r_i \frac{\partial r_i}{\partial p_k} = 0 \qquad (k = 1, n)$$

These n equations are termed the *normal equations*. To elaborate these equations further we need to derive expressions for the derivatives $\partial r_i/\partial p_k$:

$$r_i = y_i^{obs} - y_i^{calc}$$

$$\frac{\partial r_i}{\partial p_k} = -\frac{\partial y_i^{calc}}{\partial p_k}$$

For a linear model,

$$y_i^{calc} = \sum_{j=1,n} f_j(x_i)p_j \qquad (i = 1, m)$$

$f_j(x_i)$ can be any function of x_i such as a constant, x_i itself, x_i^2, $\sin x_i$, $\exp x_i$, etc. Therefore

$$\frac{\partial r_i}{\partial p_k} = -\frac{\partial y_i^{calc}}{\partial p_k} = -f_k(x_i)$$

It is the fact that these expressions are independent for the parameters that defines the model as a linear one. On substituting them and the expressions for the residuals into the normal equation, the kth normal equation becomes, after removing the factor -2,

$$\sum_{i=1,m} (y_i^{\text{obs}} - y_i^{\text{calc}}) f_k(x_i) = 0$$

$$\sum_{i=1,m} f_k(x_i) y_i^{\text{calc}} = \sum_{i=1,m} f_k(x_i) y_i^{\text{obs}}$$

$$\sum_{i=1,m} f_k(x_i) \sum_{j=1,n} f_j(x_i) p_j = \sum_{i=1,m} f_k(x_i) y_i^{\text{obs}}$$

These are n equations ($k = 1, n$) in the n unknown parameters p_j, so that by solving them those values of the parameters will be found that minimize the sum of squares S.

It is convenient to rearrange the double summation on the left-hand side of the normal equations. The kth normal equation can be written as

$$\sum_{j=1,n} \left[\sum_{i=1,m} f_k(x_i) f_j(x_i) \right] p_j = \sum_{i=1,m} f_k(x_i) y_i^{\text{obs}}$$

We can now denote the results of the inner summation, [...], by a single symbol, A_{kj}, the coefficient of the jth parameter in the kth normal equation:

$$A_{kj} = \sum_{i=1,m} f_k(x_i) f_j(x_i)$$

Likewise, the summation on the right-hand side can be written as the single symbol b_k:

$$b_k = \sum_{i=1,m} f_k(x_i) y_i^{\text{obs}}$$

The kth normal equation can now be written in a clear form:

$$\sum_{j=1,n} A_{kj} p_j = b_k$$

Let us now see how to apply these ideas in practice. We will begin with the coin mass example in its simplest form. There are no independent variables, and only one parameter, m_0, the least-squares estimate of the mass. The defining equation, or in other words the mathematical statement of the model, is therefore

$$y_i^{\text{calc}} = m_0$$

and the derivative dy_i^{calc}/dm_0 is equal to one ($f_k(x_i) = 1$ and $f_j(x_i) = 1$ for all values of j, k or i). The one normal equation becomes

$$\sum_{i=1,m} m_0 = \sum_{i=1,m} m_i$$

which has the solution

$$m_0 = \frac{\Sigma_{i=1,m}\ m_i}{m}$$

In other words, the least-squares estimate of the coin mass is simply the average of the mass measurements, and this gives some justification for using the average as a better estimate than an individual reading.

Next, let us suppose that the independent variable is time, t, and that coin mass is varying in time according to the relationship

$$y_i^{\text{calc}} = m_0 + \text{rate}\ t_i$$

m_0 and rate are the two parameters, and a graph of mass against time is going to be a straight line. The functions $f_j(x_i)$ are equal to one when j is one and t_i when j is two. We can therefore evaluate the coefficients A_{kj} as follows:

$$f_1(t_i) = 1; \qquad f_2(t_i) = t_i$$

$$A_{kj} = \sum_{i=1,m} f_k(t_i)f_j(t_i)$$

$$A_{11} = \Sigma f_1(t_i)f_1(t_i) = \Sigma 1 \times 1 = m$$
$$A_{12} = \Sigma f_1(t_i)f_2(t_i) = \Sigma 1 \times t_i = \Sigma t_i$$
$$A_{21} = \Sigma f_2(t_i)f_1(t_i) = \Sigma t_i \times 1 = \Sigma t_i$$
$$A_{22} = \Sigma f_2(t_i)f_2(t_i) = \Sigma t_i \times t_i = \Sigma t_i^2$$

The normal equations are

$$mm_0 + (\Sigma\ t_i)\ \text{rate} = \Sigma\ y_i^{\text{obs}}$$
$$(\Sigma\ t_i)m_0 + (\Sigma\ t_i^2)\ \text{rate} = \Sigma\ (t_i y_i^{\text{obs}})$$

These two equations can be solved for the two unknown parameters m_0 and rate, as discussed in Section 3.5. The result is as follows (the summation ranges are for $i = 1$ to m):

$$m_0 = \frac{(\Sigma\ t_i^2)(\Sigma\ y_i^{\text{obs}}) - (\Sigma\ t_i)\Sigma\ (t_i y_i^{\text{obs}})}{m(\Sigma\ t_i^2) - (\Sigma\ t_i)^2}$$

$$\text{Rate} = \frac{-(\Sigma\ t_i)(\Sigma\ y_i^{\text{obs}}) + m\Sigma\ (t_i y_i^{\text{obs}})}{m(\Sigma\ t_i^2) - (\Sigma\ t_i)^2}$$

3.2.2   GENERAL PROCEDURE FOR THE UNIT-WEIGHTED CASE

It is clear that once we have written down an equation defining the model we can set up and solve the normal equations to obtain values for the parameters.

The model of a linear least-squares system can be written as

$$\mathbf{y}^{\text{calc}} = \mathbf{Jp}; \qquad y_i^{\text{calc}} = \sum_{j=1,n} J_{ij} p_j$$

J stands for the *Jacobian matrix* and its elements J_{ij} are the same as $f_j(x_i)$ above. It is a rectangular array with n columns ($j = 1, n$) and m rows ($i = 1, m$):

$$J_{ij} = \frac{\partial y_i^{\text{calc}}}{\partial p_j}$$

$$\mathbf{J} = \begin{bmatrix} J_{11} & J_{12} & \cdots & J_{1n} \\ J_{21} & J_{22} & \cdots & J_{2n} \\ \cdots & \cdots & \cdots & \cdots \\ J_{m1} & J_{m2} & \cdots & J_{mn} \end{bmatrix}$$

$\mathbf{y}^{\text{calc}}$ is a vector consisting of the values y_i^{calc} ($i = 1, m$) and $\mathbf{p}$ is a vector of the parameters $p_j (j = 1, n)$. The normal equations become

$$\mathbf{J}^{\text{T}}\mathbf{Jp} = \mathbf{J}^{\text{T}}\mathbf{y}^{\text{obs}}; \qquad \sum_{j=1,n} \left(\sum_{i=1,m} J_{ik} J_{ij} \right) p_j = \sum_{i=1,m} J_{ik} y_i^{\text{obs}} \qquad (k = 1, n)$$

$\mathbf{J}^{\text{T}}$ is the *transpose* of **J**, that is $J_{ik}^{\text{T}} = J_{ki}$ for $i = 1, m$ and $k = 1, n$. $\mathbf{y}^{\text{obs}}$ is a vector consisting of the observed values y_i^{obs} ($i = 1, m$). Note that the normal equations matrix **A** is square, with n rows and columns, and is always *symmetrical*:

$$A_{jk} = \sum_{i=1,m} J_{ij} J_{ik} = \sum_{i=1,m} J_{ik} J_{ij} = A_{kj}$$

The ith row of the Jacobian contains the partial derivatives of y_i^{calc} with respect to each parameter, and it can be written as

$$[J_i] = \left[\frac{\partial y_i}{\partial p_1} \frac{\partial y_i}{\partial p_2} \cdots \frac{\partial y_i}{\partial p_n} \right]$$

The expressions in this row are all that is needed to form the normal equations, as follows. For the left-hand side, A_{kj} is the sum, over all rows, of the product of the kth and jth elements in the ith row. Similarly, the right-hand side is formed by summing, over all rows, the product of each element of $[J_i]$ with the vector of observations. In this way the normal equations can be set up without storing the whole Jacobian. First set all the elements of **A** and **b** to zero. Then, for each point i calculate $[J_i]$ and accumulate the right- and left-hand sides:

$$b_k \leftarrow b_k + [J_i]_k y_i^{\text{obs}} \qquad (k = 1, n)$$
$$A_{kj} \leftarrow A_{kj} + [J_i]_k [J_i]_j \qquad (j = 1, n; \ k = 1, j)$$

This procedure forms the upper triangle of **A**, which is all that is needed as **A** is symmetrical. For illustration we take the example of fitting the coin mass to a straight line as a function of time, t. From the definition of the model, $y_i^{\text{calc}} = m_0 + \text{rate } t_i$, it follows that

$$[J_i] = [1 \ t_i]$$

$$A_{11} = \Sigma \, 1 \times 1; \qquad A_{12} = \Sigma \, 1 \times t_i; \qquad A_{22} = \Sigma \, t_i^2$$

The saving in storage may not be trivial. For example, in an X-ray crystal structure determination there may be thousands of observations and tens of parameters; the whole Jacobian requires mn storage locations (tens of thousands) whilst one row only requires n storage locations (tens).

Note. For a linear least-squares calculation it is not necessary to have any estimates of the parameters beforehand. However, there is an alternative formulation that uses initial parameter estimates. Let these estimates be in an array $\mathbf{p}^{\text{initial}}$, and let the value of $\mathbf{y}^{\text{calc}}$ corresponding to these parameters be $\mathbf{y}^{\text{initial}}$:

$$\mathbf{y}^{\text{initial}} = \mathbf{J}\mathbf{p}^{\text{initial}}$$

After multiplying this expression on the left by $\mathbf{J}^{\text{T}}$ we obtain

$$\mathbf{J}^{\text{T}}\mathbf{y}^{\text{initial}} = \mathbf{J}^{\text{T}}\mathbf{J}\mathbf{p}^{\text{initial}}$$

Then, subtracting these equations from the normal equations $\mathbf{J}^{\text{T}}\mathbf{J}\mathbf{p} = \mathbf{J}^{\text{T}}\mathbf{y}^{\text{obs}}$ we arrive at the alternative:

$$\mathbf{J}^{\text{T}}\mathbf{J}(\mathbf{p} - \mathbf{p}^{\text{initial}}) = \mathbf{J}^{\text{T}}(\mathbf{y}^{\text{obs}} - \mathbf{y}^{\text{initial}})$$

$$\mathbf{J}^{\text{T}}\mathbf{J} \, \Delta\mathbf{p} = \mathbf{J}^{\text{T}}\mathbf{r}$$

r is a vector of residuals obtained with the initial parameter estimates. With these normal equations we calculate the correction $\Delta\mathbf{p}$ which must be added to $\mathbf{p}^{\text{initial}}$ to obtain the parameters **p**. There is a small advantage in doing this, which is that it reduces round-off errors, but this has to be balanced against the additional requirement of estimating the parameters.

3.3

3.3.1 THE WEIGHTED NORMAL EQUATIONS METHOD

In this section we shall derive the normal equations for the minimization of the weighted objective function

$$S = \sum_{i=1,m} \sum_{h=1,m} r_i W_{ih} r_h; \qquad S = \mathbf{r}^{\text{T}}\mathbf{W}\mathbf{r}$$

The weight matrix **W** should ideally be taken as the inverse of the variance–covariance matrix. However, it is also acceptable, though less useful that the weight matrix be proportional to the inverse of the variance–covariance matrix. This would imply that the relationship of one experimental error to another is known, but the absolute scale of the errors is not. In this case the proportionality constant will be estimated in the least-squares process.

Introduction of weights does not greatly complicate the least-squares calculation. By following the same procedure used in the unit-weighted case it can be shown that the kth normal equation becomes

$$\sum_{i=1,m} \sum_{h=1,m} \sum_{j=1,n} f_k(x_i) W_{ih} f_j(x_h) p_j = \sum_{i=1,m} \sum_{h=1,m} f_k(x_i) W_{ih} y_h^{obs}$$

In matrix notation the normal equations are written simply as

$$\mathbf{J}^T \mathbf{W} \mathbf{J} \mathbf{p} = \mathbf{J}^T \mathbf{W} \mathbf{y}^{obs}$$

$$\mathbf{A} \mathbf{p} = \mathbf{b}$$

where $\mathbf{A} = \mathbf{J}^T \mathbf{W} \mathbf{J}$ and $\mathbf{b} = \mathbf{J}^T \mathbf{W} \mathbf{y}^{obs}$. In the general case the weight matrix is square and has m rows and columns, and there are no short-cuts to setting up the normal equations. However, in the vast majority of data-fitting problems the weight matrix is diagonal, and the normal equations can be set up more simply.

If the experimental errors are independent of each other all covariance terms will be zero, and the variance–covariance matrix will be diagonal:

$$(COV)_{ii} = \sigma_i^2 \qquad (i = 1, m)$$

$$(COV)_{ik} = 0 \qquad (i \neq k)$$

The inverse of this diagonal matrix gives the following expression for the weights:

$$W_{ii} = 1/\sigma_i^2 \qquad (i = 1, m)$$

$$W_{ik} = 0 \qquad (i \neq k)$$

and it is clear that the weights could be stored in a vector. When the weights are diagonal the normal equations can be formed from the ith row of the Jacobian:

$$A_{kj} = \sum_{i=1,m} J_{ik} W_{ii} J_{ij}; \qquad A_{kj} \leftarrow A_{kj} + [J_i]_k W_{ii} [J_i]_j$$

$$b_k = \sum_{i=1,m} J_{ik} W_{ii} y_i^{obs}; \qquad b_k \leftarrow b_k + [J_i]_k W_{ii} y_i^{obs}$$

If the variance–covariance matrix is block-diagonal, the weight matrix will have the same structure. In this case as many rows of the Jacobian will be

needed as there are in the blocks. An example occurred in the field of potentiometric titration (see Chapter 10). When using two electrodes the weight
matrix has 2×2 blocks, so two rows of the Jacobian are needed in order to
accumulate the normal equations. If the variance–covariance matrix is not
block-diagonal the whole Jacobian has to be stored.

3.3.2   USING A FACTORED WEIGHT MATRIX

If we have a weight matrix which is symmetric, as is normally the case, it can
be decomposed into factors by a method such as Choleski decomposition:

$$\mathbf{W} = \mathbf{L}^{\mathrm{T}}\mathbf{L}$$

where $\mathbf{L}$ is an upper triangular matrix. Using this factorization (or the factorization and partial inversion of the variance–covariance matrix, as described
on p. 44) the sum of squares and the normal equations become

$$\mathbf{r}^{\mathrm{T}}\mathbf{W}\mathbf{r} \equiv \mathbf{r}^{\mathrm{T}}\mathbf{L}^{\mathrm{T}}\mathbf{L}\mathbf{r}$$
$$\mathbf{J}^{\mathrm{T}}\mathbf{W}\mathbf{J}\mathbf{p} = \mathbf{J}^{\mathrm{T}}\mathbf{W}\mathbf{y}^{\mathrm{obs}} \equiv \mathbf{J}^{\mathrm{T}}\mathbf{L}^{\mathrm{T}}\mathbf{L}\mathbf{J}\mathbf{p} = \mathbf{J}^{\mathrm{T}}\mathbf{L}^{\mathrm{T}}\mathbf{L}\mathbf{y}^{\mathrm{obs}}$$

If we make the substitutions

$$\hat{\mathbf{r}} = \mathbf{L}\mathbf{r}$$
$$\hat{\mathbf{J}} = \mathbf{L}\mathbf{J}$$
$$\hat{\mathbf{y}}^{\mathrm{obs}} = \mathbf{L}\mathbf{y}^{\mathrm{obs}}$$

the normal equations are reduced to the same form as in the unit-weighted
case:

$$\hat{\mathbf{J}}^{\mathrm{T}}\hat{\mathbf{J}}\mathbf{p} = \hat{\mathbf{J}}^{\mathrm{T}}\hat{\mathbf{y}}^{\mathrm{obs}}$$
$$S = \hat{\mathbf{r}}^{\mathrm{T}}\hat{\mathbf{r}}$$

This means that in the following discussions we do not need to include the
weights explicitly, on the understanding that $\mathbf{r}$, $\mathbf{J}$ and $\mathbf{y}$ stand for $\hat{\mathbf{r}}$, $\hat{\mathbf{J}}$ and $\hat{\mathbf{y}}$,
i.e. that the residuals, Jacobian and observations have been left-multiplied by
the weight factor matrix.[*]
   When the weights are diagonal $\mathbf{L}$ is also diagonal and each element is simply
the square root of the corresponding weight. The substitutions are then as
follows:

$$\hat{r}_i = \sqrt{W_{ii}}\, r_i$$
$$\hat{J}_{ik} = \sqrt{W_{ii}}\, J_{ik}$$
$$\hat{y}_i^{\mathrm{obs}} = \sqrt{W_{ii}}\, y_i^{\mathrm{obs}}$$

---

[*] In the alternative formulation of the normal equations, $\mathbf{J}^{\mathrm{T}}\mathbf{W}\mathbf{J}\,\Delta\mathbf{p} = \mathbf{J}^{\mathrm{T}}\mathbf{W}\mathbf{r}$, there is no need to
scale the observations separately.

$$\sum_{i=1,m} \hat{J}_{ik}\hat{J}_{ij} = \sum_{i=1,m} J_{ik}W_{ii}J_{ij}$$

$$\sum_{i=1,m} \hat{J}_{ik}\hat{y}_i^{\,\mathrm{obs}} = \sum_{i=1,m} J_{ik}W_{ii}y_i^{\,\mathrm{obs}}$$

3.4

3.4.1 THE 'RIGOROUS' LEAST-SQUARES METHOD

This section is concerned with the most general application of the principle of least squares, which has often been termed the 'rigorous' method. The essence of the general procedure is that it allows for errors to be present in both the dependent and independent variables y and x. Indeed, if both types of variable are measured experimentally, and are therefore both subject to experimental error, the distinction between dependent and independent should really disappear; we could as well say that x depends on y as the other way round. Nevertheless, we will maintain the distinction in this sense: if the model can be expressed by an equation of the form $y = f(p, x)$ then it is convenient to term y the dependent variable and x the independent one.

In the rigorous method we seek to minimize the weighted sum of squared residuals in *both* x and y. It is obvious that the residuals in x and y cannot be determined independently as there can only be one residual in each observed quantity. Therefore the residuals r_x and r_y must be related. The rigorous method of least squares obtains the relationship from the equation defining the model.

The rigorous method has been explained in some detail by Deming (1943) but his exposition is somewhat difficult to follow and lacks generality in one sense—he assumes diagonal variance–covariance matrices for the errors in x and y. Deming's exposition is an expansion of the method worked out by Gauss, the originator of the least-squares method. We shall follow Deming's treatment in broad outline, but in complete generality, and using matrix notation wherever possible in order to make the treatment clearer.

The objective is to minimize the function S:

$$S = \mathbf{r}_x^{\mathrm{T}}\mathbf{W}_x\mathbf{r}_x + \mathbf{r}_y^{\mathrm{T}}\mathbf{W}_y\mathbf{r}_y$$

$$r_{xi} = x_i^{\,\mathrm{obs}} - x_i^{\,\mathrm{calc}} = x_i^{\,\mathrm{obs}} - x_i$$

$$r_{yi} = y_i^{\,\mathrm{obs}} - y_i^{\,\mathrm{calc}} = y_i^{\,\mathrm{obs}} - y_i$$

where $\mathbf{r}_x$ and $\mathbf{r}_y$ are vectors of the residuals in x and y respectively; the superscript 'calc' is dropped from the calculated values as there is no ambiguity when the observed values are labelled. $\mathbf{W}_x$ is the weight matrix for x, and is assumed to be the inverse of the variance–covariance matrix, $\mathbf{M}_x$, of errors in x.

Similarly $\mathbf{W}_y$ is the matrix of weights for y, and is assumed to be the inverse of $\mathbf{M}_y$, the variance–covariance matrix of errors in y. The minimization is subject to the constraints that there are relationships between the two types of residual. To establish these constraints, or conditions, we define the *condition functions* $\mathcal{F}_i$:

$$\mathcal{F}_i = y_i^{\text{obs}} - f_i(r_{xi}, r_{yi}, \mathbf{p}) = 0$$

This condition function expresses the fact that when the residuals in y are subtracted from the observations, the latter are *adjusted* to be equal to the calculated values y_i^{calc}. The residuals in y, however, are not independent of the residuals in x and are dependent on the parameters, $\mathbf{p}$. The minimization is to be performed subject to these m conditions.

In order to express the relationship between the residuals as a linear relationship, it is necessary to expand the condition functions as a first-order Taylor series in the residuals. The linearization is exact in the case of linear models, but it is approximate in the case of non-linear models. The expansion is performed about some initial parameter set, preferably one close to the final one. The conditions may now be termed the *linearized conditions* as follows:

$$F_i = y_i^{\text{obs}} - y_i^{\text{calc}} - \Sigma \frac{\partial F_i}{\partial r_{xi}} r_{xi} - \Sigma \frac{\partial F_i}{\partial r_{yi}} r_{yi} - \sum_{j=1,n} \frac{\partial F_i}{\partial r_{pj}} r_{pj} = 0$$

where y_i^{calc} is the value of y for the initial parameter set. The first summation is over each of the independent variables x and becomes a single term if there is only one independent variable. Likewise the second sum becomes a single term if there is only one dependent variable. The final sum uses the 'residuals' in the parameters which are the differences between the initial and final parameter values, or parameter shifts Δp_j. Since $r_{xi} - x_i^{\text{obs}} - x_i$ and $r_{yi} = y_i^{\text{obs}} - y_i$, $r_{pj} = p_j - p_j^{\text{initial}}$,

$$K_{xhi} = \frac{\partial F_h}{\partial r_{xi}} = \frac{\partial F_h}{\partial x_i} \frac{\mathrm{d} x_i}{\mathrm{d} r_{xi}} = -\frac{\partial F_h}{\partial x_i}$$

$$K_{yhi} = \frac{\partial F_h}{\partial r_{yi}} = \frac{\partial F_h}{\partial y_i} \frac{\mathrm{d} y_i}{\mathrm{d} r_{yi}} = -\frac{\partial F_h}{\partial y_i}$$

$$J_{hj} = \frac{\partial F_h}{\partial r_{pj}} = \frac{\partial F_h}{\partial p_j} \frac{\mathrm{d} p_j}{\mathrm{d} r_{pj}} = \frac{\partial F_h}{\partial p_j}$$

We can therefore write the linearized conditions in the compact form:

$$\mathbf{F} = \Delta \mathbf{y} + \mathbf{K}_x \mathbf{r}_x + \mathbf{K}_y \mathbf{r}_y - \mathbf{J} \, \Delta \mathbf{p} = 0$$

There are m equations, one for each of the m data points.

We can now minimize S subject to the linearized conditions. To do this we use the method of undetermined multipliers.[*] This can be stated as follows. To minimize S, which is a function of $2m + n$ residuals, subject to the m linearized conditions $F_i = 0$, we must solve the $2m + n$ equations in the x and y residuals, and in the parameter residuals (parameter shifts) using the m arbitrary constants λ_h. The solution of these $2m + n$ equations will give the relationships between the variables, and the m condition equations will then give the actual parameter shifts required to minimize S:

$$\frac{\partial \Phi}{\partial r} = 0 \qquad (2m + n \text{ equations})$$

$$\Phi = S + \sum_{h=1,m} \lambda_h F_h = \mathbf{r}_x^T \mathbf{W}_x \mathbf{r}_x + \mathbf{r}_y^T \mathbf{W}_y \mathbf{r}_y + \sum_h \lambda_h F_h$$

First we differentiate with respect to an x residual:

$$\frac{\partial \Phi}{\partial r_{xi}} = 2 \sum_{l=1,m} W_{xil} r_{xl} + \sum_h \lambda_h \frac{\partial F_h}{\partial r_{xi}} = 0 \qquad (i = 1, m)$$

Now since $r_{xi} = x_i^{obs} - x_i$,

$$\frac{\partial F_h}{\partial r_{xi}} = -\frac{\partial F_h}{\partial x_i}$$

$$\frac{\partial \Phi}{\partial r_{xi}} = 2 \sum_l W_{xil} r_{xl} - \sum_h \lambda_h \frac{\partial F_h}{\partial x_i} = 0$$

There is one equation of this kind for each of the m data points. The m equations $(i = 1, m)$ can be written in matrix form as

$$2\mathbf{W}_x \mathbf{r}_x - \mathbf{K}_x^T \boldsymbol{\lambda} = 0$$

where $\boldsymbol{\lambda}$ is a vector of the undetermined multipliers. By multiplying on the left by $\mathbf{W}_x^{-1}$ we obtain an expression for the r_x residuals:

$$2\mathbf{r}_x - \mathbf{W}_x^{-1} \mathbf{K}_x^T \boldsymbol{\lambda} = 0$$

Secondly, in an analogous way we can obtain an expression for the r_y residuals. We define K_{yhi} as $\partial F_h / \partial y_i$, and express the residual as

$$2\mathbf{r}_y - \mathbf{W}_y^{-1} \mathbf{K}_y^T \boldsymbol{\lambda} = 0$$

Thirdly, we differentiate with respect to the parameter residuals; since the first two terms of Φ do not depend on the parameters,

$$\mathbf{J}^T \boldsymbol{\lambda} = 0; \qquad J_{ij} = \partial F_i / \partial p_j$$

---

[*] The term undetermined multipliers is used out of deference to Gauss. In modern terminology they would be called Lagrange multipliers.

We now have four sets of equations:

$$2\mathbf{r}_x = \mathbf{M}_x\mathbf{K}_x^T\boldsymbol{\lambda} \qquad (\mathbf{M}_x = \mathbf{W}_x^{-1})$$

$$2\mathbf{r}_y = \mathbf{M}_y\mathbf{K}_y^T\boldsymbol{\lambda} \qquad (\mathbf{M}_y = \mathbf{W}_y^{-1})$$

$$\mathbf{J}^T\boldsymbol{\lambda} = 0$$

$$\mathbf{F} = \Delta\mathbf{y} + \mathbf{K}_x\mathbf{r}_x + \mathbf{K}_y\mathbf{r}_y - \mathbf{J}\;\Delta\mathbf{p} = 0$$

We may substitute the expressions for the residuals into the condition equations, with the following result:

$$\Delta\mathbf{y} + \tfrac{1}{2}\mathbf{K}_x\mathbf{M}_x\mathbf{K}_x^T\boldsymbol{\lambda} + \tfrac{1}{2}\mathbf{K}_y\mathbf{M}_y\mathbf{K}_y^T\boldsymbol{\lambda} - \mathbf{J}\;\Delta\mathbf{p} = 0$$

To simplify the notation we define $\mathbf{M}$ as

$$\mathbf{M} = \mathbf{K}_x\mathbf{M}_x\mathbf{K}_x^T + \mathbf{K}_y\mathbf{M}_y\mathbf{K}_y^T$$

therefore

$$\Delta\mathbf{y} + \tfrac{1}{2}\mathbf{M}\boldsymbol{\lambda} - \mathbf{J}\;\Delta\mathbf{p} = 0$$

$$\boldsymbol{\lambda} = -2\mathbf{M}^{-1}\Delta\mathbf{y} + 2\mathbf{M}^{-1}\mathbf{J}\;\Delta\mathbf{p}$$

Finally, this expression for the multipliers, which are now determined, is substituted into the expression $\mathbf{J}^T\boldsymbol{\lambda} = 0$ to reveal the solution:

$$\mathbf{J}^T\mathbf{M}^{-1}\mathbf{J}\;\Delta\mathbf{p} = \mathbf{J}^T\mathbf{M}^{-1}\;\Delta\mathbf{y}$$

The result of all this labour is reassuringly familiar. We recognize the equations as the standard equations for non-linear least squares, with $\mathbf{M}$ clearly identified with a variance–covariance matrix. Therefore the final message is that to take into consideration errors on both x and y all that is required is to construct the correct variance–covariance matrix from which to derive, by inverting the matrix, the weights for the observations in the dependent variable y. These weights depend on errors in both x and y. However, before giving some examples to clarify the procedure we should perhaps look in more detail at the relationship between the x and y residuals:

$$\mathbf{r}_x = \mathbf{M}_x\mathbf{K}_x^T\boldsymbol{\lambda}, \qquad \mathbf{r}_y = \mathbf{M}_y\mathbf{K}_y^T\boldsymbol{\lambda}$$

We can see the relationship most clearly if we make the following simplifying assumptions, which apply in the most common case: the variables x_i and y_i only affect the condition function F_i, which makes $\mathbf{K}_x$ and $\mathbf{K}_y$ diagonal, and errors are uncorrelated, so that $\mathbf{M}_x$ and $\mathbf{M}_y$ are also diagonal. Then

$$r_{xi} = M_{xii}\,\partial F_i/\partial x_i\;\lambda_i$$

$$r_{yi} = M_{yii}\,\partial F_i/\partial y_i\;\lambda_i$$

$$\frac{r_{yi}}{r_{xi}} = \frac{M_{yii}\,\partial F_i/\partial y_i\;\lambda_i}{M_{xii}\,\partial F_i/\partial x_i\;\lambda_i} = -\frac{M_{yii}}{M_{xii}}\frac{\mathrm{d}y}{\mathrm{d}x}$$

This shows that the ratio of the residuals on y and x at the ith point is proportional to minus the slope of the curve of y against x at that point, the proportionality constant being the ratio of the variances. This becomes very clear if the ratio of the variances is unity and the x and y scales are the same. In that case the line between an observed and the corresponding calculated point is perpendicular to the tangent of the curve of y against x. This makes good intuitive sense, since, when the errors on x and y are the same, we would expect to be minimizing the perpendicular distances between the calculated and observed curves, just as, when x is free from error, we minimize the vertical distance (taking the y axis as vertical). Although we can calculate the residuals on x and y separately, it makes little sense to do this since they are interrelated.

3.4.2 CALCULATION OF THE VARIANCE–COVARIANCE MATRIX

A number of different cases can be considered. In each case we will need the partial derivatives of the condition function with respect to y, K_y:

$$F_i = \Delta \mathbf{y} + \mathbf{K}_x \mathbf{r}_x + \mathbf{K}_y \mathbf{r}_y - \mathbf{J} \, \Delta \mathbf{p} = 0 \qquad (\Delta \mathbf{y} = \mathbf{y}^{\text{obs}} - \mathbf{y})$$

If we assume that any one y, y_i say, enters into only one condition, F_i, then K_y is diagonal with $K_{ii} = -1$. $\mathbf{K}_y$ is therefore -1 times an identity matrix, and $\mathbf{K}_y \mathbf{M}_y \mathbf{K}_y^T$ is equal to $\mathbf{M}_y$. This assumption is justified in the overwhelming majority of least-squares calculations.

(a) There are no errors on x. In this case $\mathbf{M}_x$ is a null matrix and, assuming that $\mathbf{K}_y = -\mathbf{I}$, $\mathbf{M} = \mathbf{M}_y$, the variance–covariance matrix of the observations.

(b) There is one independent variable x, subject to error, and one dependent variably y, also subject to error. This is by far the most common situation. For example, the model for fitting a straight line is given by

$$f(x_i, \mathbf{p}) = p + q x_i; \qquad K_{xii} = -q$$

Assuming that $\mathbf{K}_y$ is minus an identity matrix, the $\mathbf{M}$ matrix becomes

$$\mathbf{M} = \mathbf{K}_y \mathbf{M}_y \mathbf{K}_y^T + \mathbf{K}_x \mathbf{M}_x \mathbf{K}_x^T = \mathbf{M}_y + q^2 \mathbf{M}_x$$

When the errors are uncorrelated and both $\mathbf{M}_x$ and $\mathbf{M}_y$ are diagonal, the variance–covariance matrix $\mathbf{M}$ becomes

$$M_{ii} = \sigma_{yi}^2 + q^2 \sigma_{xi}^2$$

This same formula will be obtained again in Section 4.1 from a consideration of error propagation. In fact this formula is quite useful for general non-linear functions which can be expanded as a first-order Taylor series in x:

$$y = f(x_i, \mathbf{p}) = y^0 + (\partial y/\partial x_i)x_i$$
$$M_{ii} = \sigma_{yi}^2 + (\partial y/\partial x_i)^2 \sigma_{xi}^2$$

Rather than calculate the slope of the curve with parameter estimates it is usually acceptable to use the slope of the *observed* y against x.

For another example consider the fitting of a parabola:

$$f(x_i, \mathbf{p}) = p + qx_i + rx_i^2$$
$$K_{xii} = -q - 2rx_i$$
$$M_{ij} = M_{yij} + (q + 2rx_i)(q + 2rx_j)M_{xij}$$

This example highlights a point that needs some emphasis. In the rigorous method of least squares we always begin with some parameter estimates and proceed to calculate the parameter shift required to minimize the sum of squares. This is as true for linear models (like the parabola above) as for non-linear models. It is important that the parameter shifts should be small so that observed and calculated curves nearly coincide in order that the correct $\mathbf{M}$ matrix may be calculated. If need be, one must do the calculation once with one set of parameter estimates, and then again with the optimal set.

(c) There are two observations for each independent variable, all three subject to experimental error. This situation arises whenever two properties are measured simultaneously. There is good reason to do such experiments, especially if the two properties give complementary information. An example may occur in the field of solution equilibria. Normally the equilibrium constants are obtained from measurements of the pH of a solution as a function of added alkali. Alternatively, if one or more of the species in equilibrium is coloured it may be followed by measuring optical absorbance, A. To utilize both sets of information each measurement really does need to be given its correct weight.

We may assume that the errors on these two measurements are uncorrelated as they are supposed to be independent. To take the example just discussed, let us consider the data at the first point, where the two observations depend on the first value of the independent variable:

$$F_1 = A - f_1(x_1)$$
$$F_2 = \mathrm{pH} - f_2(x_1)$$

Both absorbance and pH depend on the same volume of alkali, which we have designated as x_1. Because each observation occurs in only one condition equation, $\mathbf{K}_y$ is minus the identity matrix, and the first term in $\mathbf{M}$ is therefore simply $\mathbf{M}_y$. For the $\mathbf{K}_x$ matrix we require the two derivatives:

$$K_{A,x1} = \frac{\partial F_1}{\partial x_1}, \qquad K_{pH,x1} = \frac{\partial F_2}{\partial x_1}$$

Assuming for the moment that errors in volume are not correlated, $\mathbf{M}_x$ is diagonal and the second term in $\mathbf{M}$ is given by

$$\mathbf{K}_x \mathbf{M}_x \mathbf{K}_x^T = \begin{bmatrix} \partial A/\partial x \\ \partial pH/\partial x \end{bmatrix} \sigma_x^2 [\partial A/\partial x \; \partial pH/\partial x]$$

$$= \begin{bmatrix} (\partial A/\partial x)^2 \sigma_x^2 & (\partial A/\partial x)(\partial pH/\partial x)\sigma_x^2 \\ (\partial A/\partial x)(\partial pH/\partial x)\sigma_x^2 & (\partial pH/\partial x)^2 \sigma_x^2 \end{bmatrix}$$

Here we have used the relationships

$$\partial F_1/\partial x \approx \partial A/\partial x, \qquad \partial F_2/\partial x \approx \partial pH/\partial x$$

which are the slopes of the observed values of absorbance and pH as a function of x. Adding together $\mathbf{K}_x$ and $\mathbf{K}_y$ we obtain $\mathbf{M}$:

$$\mathbf{M} = \begin{bmatrix} \sigma_A^2 + (\partial A/\partial x)^2 \sigma_x^2 & (\partial A/\partial x)(\partial pH/\partial x)\sigma_x^2 \\ (\partial A/\partial x)(\partial pH/\partial x)\sigma_x^2 & \sigma_{pH}^2 + (\partial pH/\partial x)^2 \sigma_x^2 \end{bmatrix}$$

This is the part of the $\mathbf{M}$ matrix that refers to the first pair of observations. Similar 2×2 blocks appear on the diagonal of $\mathbf{M}$ for the other pairs of observations. If $\mathbf{M}_y$ is diagonal the whole $\mathbf{M}$ matrix is block-diagonal and the inversion of the 2×2 blocks is straightforward:

$$M_{11}^{-1} = M_{22}/D; \quad M_{12}^{-1} = -M_{12}/D; \quad M_{22}^{-1} = M_{11}/D; \quad D = M_{11}M_{22} - M_{12}^2$$

Notice that even though we have assumed that both errors on x and errors on y were uncorrelated, there is a covariance term because both observations depend on the same independent variable.

(d) There are two independent variables for each measurement, all three variables subject to error. An example might be drawn from chemical kinetics where a pseudo first-order reaction rate was measured at various temperatures, T, and concentrations, c, of an excess reactant. The observed rate constant, y, derived from the observed reaction rates as a function of time, depends on the two independent variables c and T, and is a function of the two parameters known as Arrhenius parameters:

$$F_1 = y^{obs} - f(c_1, T_1, \mathbf{p})$$

In this case we shall consider in detail only the first observation. The partial derivative with respect to y is -1, so we can see that

$K_y M_y K_y^T = M_y$, as usual. For K_x we must consider the derivatives with respect to the two independent variables:

$$K_x M_x K_x^T = [\partial y/\partial c \ \ \partial y/\partial T] \begin{bmatrix} \sigma_c^2 & 0 \\ 0 & \sigma_T^2 \end{bmatrix} \begin{bmatrix} \partial y/\partial c \\ \partial y/\partial T \end{bmatrix} = (\partial y/\partial c)^2 \sigma_c^2 + (\partial y/\partial T)^2 \sigma_T^2$$

$$M_{11} = \sigma_{y1}^2 + (\partial y/\partial c)_1^2 \sigma_{c1}^2 + (\partial y/\partial T)_1^2 \sigma_{T1}^2$$

assuming that M_y, M_c and M_T are diagonal. Once again the form of M for the other data points is the same so in this case M is diagonal.

3.4.3 THE GENERAL VARIANCE–COVARIANCE MATRIX

We seek to form the variance–covariance matrix:

$$M = K_y M_y K_y^T + K_x M_x K_x^T$$

where M_x and M_y are the variance–covariance matrices for the x and y variables. Both K_x and K_y matrices are blocked; the size of each block depends on the number of dependent and independent variables that enter into each condition function. Consider the first set of data points. The structure of the first block of K_y is as follows:

$$K_y = \begin{bmatrix} \dfrac{\partial F_1}{\partial y_1} & \dfrac{\partial F_1}{\partial y_2} & \cdots & \dfrac{\partial F_1}{\partial y_k} \\[2ex] \dfrac{\partial F_2}{\partial y_1} & \dfrac{\partial F_2}{\partial y_2} & \cdots & \dfrac{\partial F_2}{\partial y_k} \\[2ex] \cdots & \cdots & \cdots & \cdots \\[2ex] \dfrac{\partial K_j}{\partial y_1} & \dfrac{\partial F_j}{\partial y_2} & \cdots & \dfrac{\partial F_j}{\partial y_k} \end{bmatrix}$$

This structure shows that there are k dependent variables that occur in the first j condition equations. All other elements of K_y are zero in these rows and columns. The next block begins at row $j + 1$ and column $k + 1$, and has the same structure, with the second set of condition functions and dependent variables in place of the first, and so on.

The structure of the K_x matrix is similar, each block having j' rows and k' columns where the first k' independent variables occur in the first j' condition equations. The block structure in the three cases outlined above is then as follows:

(b) One dependent and one independent variable. Both K_x and K_y have the same structure, and can be represented by the diagonal matrices K. M is understood to refer to either $\mathbf{M}_x$ or $\mathbf{M}_y$:

$$\mathbf{K} = \begin{bmatrix} K_{11} & 0 & 0 & \cdots \\ 0 & K_{22} & 0 & \cdots \\ 0 & 0 & K_{33} & \cdots \\ \cdots & \cdots & \cdots & \cdots \end{bmatrix}; \qquad \mathbf{M} = \begin{bmatrix} M_{11} & M_{12} & M_{13} & \cdots \\ M_{12} & M_{22} & M_{23} & \cdots \\ M_{13} & M_{23} & M_{33} & \cdots \\ \cdots & \cdots & \cdots & \cdots \end{bmatrix}$$

Having taken both $\mathbf{K}_x$ and $\mathbf{K}_y$ to be diagonal implies that each dependent and independent variable appears in one condition equation only. When this is so $\mathbf{K}_y$ is a negative identity matrix and $\mathbf{K}_y \mathbf{M}_y \mathbf{K}_y^T = \mathbf{M}_y$. $\mathbf{K}_x$ is diagonal, so

$$(\mathbf{K}_x \mathbf{M}_x \mathbf{K}_x^T)_{gh} = K_{xgg} M_{xgh} K_{xhh}$$

$$M_{gh} = M_{ygh} + K_{xgg} M_{xgh} K_{xhh}$$

(c) Two observations for each independent variable.

$$\mathbf{K}_y = -\mathbf{I}$$

Because each pair of observations depends on the same independent variable $\mathbf{K}_x$ has the form

$$\mathbf{K}_x = \begin{bmatrix} K_{11} & 0 & 0 & \cdots \\ K_{21} & 0 & 0 & \cdots \\ 0 & K_{32} & 0 & \cdots \\ 0 & K_{42} & 0 & \cdots \\ 0 & 0 & K_{53} & \cdots \\ 0 & 0 & K_{63} & \cdots \\ \cdots & \cdots & \cdots & \cdots \end{bmatrix}$$

The odd-numbered rows refer to the first observation, whilst the even-numbered rows refer to the second observation. It is convenient to group the former and latter together, by rearranging the rows:

$$\mathbf{K}_x = \left[\begin{array}{ccccc} K_{11} & 0 & 0 & \cdots \\ 0 & K_{32} & 0 & \cdots \\ 0 & 0 & K_{53} & \cdots \\ \cdots & \cdots & \cdots & \cdots \\ \hline K_{21} & 0 & 0 & \cdots \\ 0 & K_{42} & 0 & \cdots \\ 0 & 0 & K_{63} & \cdots \\ \cdots & \cdots & \cdots & \cdots \end{array} \right] = \begin{pmatrix} \mathbf{K}_1 \\ \mathbf{K}_2 \end{pmatrix}$$

Each subblock $\mathbf{K}_1$ and $\mathbf{K}_2$ now has $m/2$ rows and m columns. The $\mathbf{M}_x$

matrix has m rows and columns. Therefore the required result is

$$\mathbf{K}_x\mathbf{M}_x\mathbf{K}_x^\mathsf{T} = \begin{pmatrix} \mathbf{K}_1 \\ \mathbf{K}_2 \end{pmatrix} \mathbf{M}_x (\mathbf{K}_1^\mathsf{T} \mid \mathbf{K}_2^\mathsf{T}) = \begin{pmatrix} \mathbf{K}_1\mathbf{M}_x\mathbf{K}_1^\mathsf{T} \mid \mathbf{K}_1\mathbf{M}_x\mathbf{K}_2^\mathsf{T} \\ \mathbf{K}_2\mathbf{M}_x\mathbf{K}_1^\mathsf{T} \mid \mathbf{K}_2\mathbf{M}_x\mathbf{K}_2^\mathsf{T} \end{pmatrix}$$

which has m rows and columns. Rearrangement of the rows into their original order and a corresponding rearrangement of the columns gives

$$(\mathbf{K}_x\mathbf{M}_x\mathbf{K}_x^\mathsf{T})_{gh} = K_{xgi}M_{xij}K_{xhj}$$

The single term results because there is only one non-zero element in each row, g and h, of $\mathbf{K}_x$. It will be readily seen that if $\mathbf{M}_x$ is diagonal, $\mathbf{M}$ will be 2×2 block-diagonal, as discussed above.

(d) Two independent variables for each observation. $\mathbf{K}_y$ will normally be a unit matrix. We could write the $\mathbf{K}_x$ matrix as

$$\mathbf{K}_x = \begin{bmatrix} K_{11} & K_{12} & 0 & 0 & 0 & 0 & \cdots \\ 0 & 0 & K_{23} & K_{24} & 0 & 0 & \cdots \\ 0 & 0 & 0 & 0 & K_{35} & K_{36} & \cdots \\ \cdots & \cdots & \cdots & \cdots & \cdots & \cdots & \cdots \end{bmatrix}$$

However, the odd-numbered columns refer to the first independent variable and the even-numbered columns refer to the second independent variable. It is therefore more convenient to rearrange the columns so that the derivatives for the first variable appear in one block, and those for the second in a separate block:

$$\mathbf{K}_x = \left[\begin{array}{cccc|cccc} K_{11} & 0 & 0 & \cdots & K_{12} & 0 & 0 & \cdots \\ 0 & K_{23} & 0 & \cdots & 0 & K_{24} & 0 & \cdots \\ 0 & 0 & K_{35} & \cdots & 0 & 0 & K_{36} & \cdots \\ \cdots & \cdots & \cdots & \cdots & \cdots & \cdots & \cdots & \cdots \end{array} \right] = (\mathbf{K}_1 \mid \mathbf{K}_2)$$

The two variance–covariance matrices can now be placed as blocks in the variance–covariance matrix giving the errors in the two variables. We can usually assume that the errors on the two kinds of independent variable are uncorrelated:

$$\mathbf{M}_x = \begin{bmatrix} \mathbf{M} & 0 \\ 0 & \mathbf{N} \end{bmatrix}$$

$$\mathbf{K}_x\mathbf{M}_x^\mathsf{T}\mathbf{K}_x = [\mathbf{K}_1 \mid \mathbf{K}_2] \begin{bmatrix} \mathbf{M} & 0 \\ 0 & \mathbf{N} \end{bmatrix} \begin{pmatrix} \mathbf{K}_1^\mathsf{T} \\ \mathbf{K}_2^\mathsf{T} \end{pmatrix}$$

$$= \mathbf{K}_1\mathbf{M}\mathbf{K}_1^\mathsf{T} + \mathbf{K}_2\mathbf{N}\mathbf{K}_2^\mathsf{T}$$

After rearranging the columns back to their original order, we obtain

$$(\mathbf{K}_x\mathbf{M}_x\mathbf{K}_x^\mathsf{T})_{gh} = K_{xgg}M_{xgh}K_{xhh} + K_{xg'g'}N_{xg'h'}K_{xh'h'}$$

where $g' = g + 1$ and $h' = h + 1$.

These details have been given for the sake of completeness. In the vast majority of data-fitting problems it can usually be arranged for the errors to be uncorrelated. Although this brings some simplification to the algebra it is not necessary, and uncorrelated errors need not be assumed when there is evidence to the contrary.

Finally, let us observe that the usual way of posing the least-squares problem, in which the independent variable is free from error, can be derived from the general case simply by putting $\mathbf{M}_x = 0$. In this way the case of two observations per independent variable, or two variables per observation, can be easily treated by using the formulae given above. Also the very unlikely case where one dependent variable enters more than one condition equation relating to the same kind of observation can be derived by the general procedure. Deming mentions such a case, where one observation depends directly on a previous observation (not just that their errors are correlated).

3.4.4   PARAMETER-FREE LEAST SQUARES

This is the original application developed by Gauss. Models without parameters arise when there is a relationship between observations that does not depend on parameters. The simplest example is the determination of the angles in a triangle. Suppose the three angles have been measured and the sum is not exactly 180 degrees. The question is, how should the angles be best adjusted?

We solve this problem by writing down the one condition equation:

$$180 - \alpha - \beta - \gamma - \delta\alpha - \delta\beta - \delta\gamma = 0$$

where α, β and γ are the measured angles and $\delta\alpha$, $\delta\beta$ and $\delta\gamma$ are the adjustments required to make the sum of the angles equal to 180. We seek to minimize the weighted sum of square adjustments:

$$S = W_1\,\delta\alpha^2 + W_2\,\delta\beta^2 + W_3\,\delta\gamma^2$$

Following the general procedure in Section 3.4.1, the method of undetermined multipliers yields three equations:

$$2\mathbf{r} = \mathbf{MK\lambda}$$

As we have posed the problem, the weight matrix is diagonal and $\mathbf{K}$ is a vector whose elements are all -1. Writing $\Delta = 180 - \alpha - \beta - \gamma$, the ensemble of all equations is

$$\Delta - \delta\alpha - \delta\beta - \delta\gamma = 0$$
$$2\delta\alpha = -M_1\lambda$$
$$2\delta\beta = -M_2\lambda$$
$$2\delta\gamma = -M_3\lambda$$

The solution to these four equations is

$$\delta\alpha = M_1 \Delta / (M_1 + M_2 + M_3)$$
$$\delta\beta = M_2 \Delta / (M_1 + M_2 + M_3)$$
$$\delta\gamma = M_3 \Delta / (M_1 + M_2 + M_3)$$

If the weights are equal, the adjustment to each angle is simply one third of the discrepancy Δ. This result is not just of historical interest. It is routinely used by astronomers, for whom it was derived. It shows how the method of least squares does not necessarily involve parameters. Another example might be a total chemical analysis of a compound. If the individual element percentages do not add up to 100%, they should be adjusted by the above method.

3.5

3.5.1 SOLVING THE NORMAL EQUATIONS

The normal equations are simply n linear equations in the n unknown parameters, and there are many ways in which they can be solved numerically. When there is only one parameter the equation and its solution are

$$Ap = b; \qquad p = b/A$$

The constants A and b are calculated by making the appropriate summations as indicated in Sections 3.2 and 3.3. With two parameters the normal equations are

$$A_{11}p_1 + A_{12}p_2 = b_1$$
$$A_{21}p_1 + A_{22}p_2 = b_2$$

One way to solve these equations is to eliminate one variable as follows. Multiply the first equation by A_{22} and the second equation by A_{12}:

$$A_{11}A_{22}p_1 + A_{12}A_{22}p_2 + A_{22}b_1$$
$$A_{12}A_{21}p_1 + A_{12}A_{22}p_2 = A_{12}b_2$$

On subtracting the second equation from the first we obtain

$$A_{11}A_{22}p_1 - A_{12}A_{21}p_1 = A_{22}b_1 - A_{12}b_2$$
$$(A_{11}A_{22} - A_{12}A_{21})p_1 = A_{22}b_1 - A_{12}b_2$$

Using the fact that $A_{12} = A_{21}$ we can write

$$D = A_{11}A_{22} - A_{12}^2$$
$$p_1 = (A_{22}/D)b_1 - (A_{12}/D)b_2$$

In a similar way we obtain

$$p_2 = -(A_{21}/D)b_1 + (A_{11}/D)b_2$$

While this method of solution is adequate for the case of two parameters and can be extended for the case of three parameters, it is not the most convenient method for general use. It does, however, reveal some important characteristics of the general solution. The coefficients of b_1 and b_2 are elements of the matrix known as the inverse of the matrix $\mathbf{A}$ of coefficients of p_1 and p_2 in the normal equations (see Appendix 2 for a definition of the inverse of a matrix). If we denote A_{22}/D as C_{11}, $-A_{12}/D$ as C_{12} and so forth, we can write

$$\mathbf{Ap} = \mathbf{b}$$

$$\mathbf{p} = \mathbf{Cb}$$

where $\mathbf{C} = \mathbf{A}^{-1}$, which has the meaning

$$\sum_{k=1,n} A_{ik}C_{ki} = \sum_{k=1,n} C_{ik}A_{ki} = 1$$

$$\sum_{k=1,n} A_{ik}C_{kj} = \sum_{k=1,n} C_{ik}A_{kj} = 0 \qquad (i \neq j)$$

The general solution of the normal equations can be written as

$$\mathbf{p} = (\mathbf{J}^T\mathbf{WJ})^{-1}\mathbf{J}^T\mathbf{Wy}^{\text{obs}}$$

Again, this expression does not represent the most convenient way of solving the normal equations, but it does give the solution in a very compact form. The inverse of the normal equations matrix also plays an important role in estimating the errors in the calculated parameters, as we shall demonstrate in Section 3.6.

3.5.2 GENERAL METHOD FOR SOLVING THE NORMAL EQUATIONS

The recommended method to use to solve simultaneous equations when the coefficients form a symmetric matrix is a computer program based on a form of the Choleski method, which has three stages. Firstly, the coefficients matrix is factored:

$$\mathbf{A} = \mathbf{LL}^T$$

Then the solution is found by forward and backward substitutions:

$$\mathbf{Lz} = \mathbf{b}$$

$$\mathbf{L}^T\mathbf{p} = \mathbf{z}$$

Together these stages are equivalent to solving $\mathbf{Ap} = \mathbf{b}$.

The factor matrix **L** is lower triangular:

$$\mathbf{L} = \begin{bmatrix} L_{11} & 0 & 0 & \cdots & 0 \\ L_{21} & L_{22} & 0 & \cdots & 0 \\ L_{31} & L_{32} & L_{33} & \cdots & 0 \\ \cdots & \cdots & \cdots & \cdots & \cdots \\ L_{n1} & L_{n2} & L_{n3} & \cdots & L_{nn} \end{bmatrix}, \qquad \mathbf{L}^{\mathrm{T}} = \begin{bmatrix} L_{11} & L_{21} & L_{31} & \cdots & L_{n1} \\ 0 & L_{22} & L_{32} & \cdots & L_{n2} \\ 0 & 0 & L_{33} & \cdots & L_{n3} \\ \cdots & \cdots & \cdots & \cdots & \cdots \\ 0 & 0 & 0 & \cdots & L_{nn} \end{bmatrix}$$

The elements can be evaluated in turn by going across the rows, starting at the first, and working downwards:

$$L_{11}L_{11} = A_{11} \qquad L_{11} = \sqrt{A_{11}}$$
$$L_{21}L_{11} = A_{12} \qquad L_{21} = A_{12}/L_{11}$$
$$L_{21}L_{21} + L_{22}L_{22} = A_{22} \qquad L_{22} = \sqrt{(A_{22} - L_{21}^2)}$$
$$\cdots$$

The sequence continues $L_{31} \cdots L_{33}, \ldots, L_{n1} \cdots L_{nn}$. The calculation of each element involves only elements previously calculated, so **L** may overwrite the lower triangle of **A** and only the diagonal elements of **A** need to be stored separately so as not to be lost. It should be noted that this method requires the taking of square roots. If **A** is positive definite all the quantities concerned should be positive; if one is not positive the factorization fails. Also, the product for the diagonal elements of **L** is equal to the square root of the determinant of **A**.

The forward substitution also follows a sequence such that each new element of z depends only on previously calculated elements, so z may overwrite the right-hand side vector **b**:

$$L_{11}z_1 = b_1 \qquad z_1 = b_1/L_{11}$$
$$L_{21}z_1 + L_{22}z_2 = b_2 \qquad z_2 = (b_2 - L_{21}z_1)/L_{22}$$
$$L_{31}z_1 + L_{32}z_2 + L_{33}z_3 = b_3 \qquad z_3 = (b_3 - L_{31}z_1 - L_{32}z_2)/L_{22}$$
$$\cdots$$

The backward substitution works in a similar way, but starts with the last element and works upwards, with **p** overwriting **b** again:

$$L_{nn}p_n = z_n$$
$$p_n = z_n/L_{nn}$$
$$L_{n-1,n-1}p_{n-1} + L_{n-1,n}p_n = z_{n-1}$$
$$p_{n-1} = (z_{n-1} - L_{n-1,n}z_n)/L_{n-1,n-1}$$
$$p_{n-2} = (z_{n-2} - L_{n-2,n-1}z_{n-1} - L_{n-2,n}p_n)/L_{n-2,n-2}$$
$$\cdots$$

The normal equations matrix $\mathbf{A}$ can be inverted using the factor $\mathbf{L}$:

$$\mathbf{A}^{-1} = (\mathbf{LL}^{T})^{-1} = (\mathbf{L}^{T})^{-1}\mathbf{L}^{-1} = (\mathbf{L}^{-1})^{T}\mathbf{L}^{-1} = \mathbf{Z}^{T}\mathbf{Z}$$

$$\mathbf{LZ} = \mathbf{I}$$

The columns of $\mathbf{Z}$ are obtained by forward and backward substitutions on columns of the identity matrix $\mathbf{I}$. $\mathbf{Z}$ has the same lower triangular form as $\mathbf{L}$.

An alternative form of Choleski factorization uses the identity $\mathbf{A} = \mathbf{LDL}^{T}$ and the substitutions $\mathbf{Lz} = \mathbf{b}$ and $\mathbf{DL}^{T}\mathbf{p} = \mathbf{z}$. $\mathbf{D}$ is a diagonal matrix and $\mathbf{L}$ has the form

$$\mathbf{L} = \begin{bmatrix} 1 & 0 & 0 & \cdots & 0 \\ L_{21} & 1 & 0 & \cdots & 0 \\ L_{31} & L_{32} & 1 & \cdots & 0 \\ \cdots & \cdots & \cdots & \cdots & \cdots \\ L_{n1} & L_{n2} & L_{n3} & \cdots & 1 \end{bmatrix}, \qquad \mathbf{L}^{T} = \begin{bmatrix} 1 & L_{21} & L_{31} & \cdots & L_{n1} \\ 0 & 1 & L_{32} & \cdots & L_{n2} \\ 0 & 0 & 1 & \cdots & L_{n3} \\ \cdots & \cdots & \cdots & \cdots & \cdots \\ 0 & 0 & 0 & \cdots & 1 \end{bmatrix}$$

The main difference in using this form is that there is no requirement to take square roots. It is therefore more convenient when using Newton's method (Section 4.7.1) with a symmetric coefficients matrix. I personally prefer this form for general use also. The upper triangle for $\mathbf{A}$ remains unchanged as there is no need to store the unit diagonal elements of $\mathbf{L}$ explicitly. The determinant of $\mathbf{A}$ is the product of the elements of $\mathbf{D}$, which is, of course, stored in a vector rather than a two-dimensional array.

Note. The Choleski method can also be applied to the factorization of a weight matrix as long as it is symmetrical. In fact, if the variance–covariance matrix of the observations is not sparse it may be more convenient to use the Choleski method to factor $\mathbf{M}$ and obtain the inverse factors directly without actually calculating the inverse of $\mathbf{M}$:

$$\mathbf{M} = \mathbf{LL}^{T}$$

$$\mathbf{LZ} = \mathbf{I}$$

$$\mathbf{Z} = \mathbf{L}^{-1}$$

$$\mathbf{W} = \mathbf{M}^{-1} = (\mathbf{Z}^{T})^{-1}\mathbf{Z}$$

$$\hat{\mathbf{r}} = \mathbf{Zr}$$

$$\hat{\mathbf{J}} = \mathbf{ZJ}$$

$$\hat{\mathbf{y}}^{\text{obs}} = \mathbf{Zy}^{\text{obs}}$$

3.6

3.6.1   PARAMETER STANDARD DEVIATIONS AND CORRELATION COEFFICIENTS

The theory of error propagation (Section 2.2) shows that when one set of variables is a linear combination of another set

$$\mathbf{p} = \mathbf{Ty}$$

the variance–covariance matrix, $\mathbf{P}$, on the set $\mathbf{p}$ is given by

$$\mathbf{P} = \mathbf{TMT}^{\mathrm{T}}$$

where $\mathbf{M}$ is the variance–covariance matrix on $\mathbf{y}$.

The parameters obtained by the method for linear least squares are linear combinations of the observations, with the linear transformation coefficients given by the normal equations (Section 3.5):

$$\mathbf{p} = (\mathbf{J}^{\mathrm{T}}\mathbf{WJ})^{-1}\mathbf{J}^{\mathrm{T}}\mathbf{Wy}^{\mathrm{obs}} = \mathbf{Ty}^{\mathrm{obs}}$$
$$\mathbf{T} = (\mathbf{J}^{\mathrm{T}}\mathbf{WJ})^{-1}\mathbf{J}^{\mathrm{T}}\mathbf{W}$$

Therefore the variance–covariance matrix of the parameters is given by

$$\mathbf{P} = (\mathbf{J}^{\mathrm{T}}\mathbf{WJ})^{-1}\mathbf{J}^{\mathrm{T}}\mathbf{WMW}^{\mathrm{T}}\mathbf{J}(\mathbf{J}^{\mathrm{T}}\mathbf{WJ})^{-1}$$

(We know that the normal equations matrix and its inverse are symmetrical so that the transpose of this matrix is the matrix itself.) This equation can be greatly simplified if we allow the weight matrix $\mathbf{W}$ to be the inverse of the symmetric matrix $\mathbf{M}$. In that case the product $\mathbf{MW}^{\mathrm{T}}$ is equal to the identity and can be removed from the equation, yielding

$$\mathbf{P} = (\mathbf{J}^{\mathrm{T}}\mathbf{WJ})^{-1}\mathbf{J}^{\mathrm{T}}\mathbf{WJ}(\mathbf{J}^{\mathrm{T}}\mathbf{WJ})^{-1} = (\mathbf{J}^{\mathrm{T}}\mathbf{WJ})^{-1}$$

Thus the required result, the variance–covariance matrix of parameters, is simply the inverse of the normal equations matrix. Notice that we have introduced one assumption into this treatment, namely that the weight matrix should be the inverse of the variance–covariance matrix of the observations. This is not a very restrictive assumption, but it should be remembered that the assumption has been made.

There is an interesting implication in this important result. The correlation coefficients between parameters are in general non-zero, even when there is no correlation between the observations. We can also see the value of giving the observations their correct weights in the least-squares calculation. It does not matter if the observations are uncorrelated and have equal errors; in that case it would be better to give each observation a weight equal to the reciprocal of the variance than to give each observation unit weight, as is often done. Now that we are using electronic computers to do the arithmetic there is no need to reduce the amount of numerical work by ignoring the weights.

Let us now work out an example, to see that the process of obtaining the errors on the parameters involves only a small amount of extra work after obtaining the values of the parameters themselves. In order to do the latter we have to set up and solve the normal equations. It therefore only remains to invert the normal equations matrix and print out the results. Take the case of straight line fitting with uncorrelated errors and each observation y_i^{obs} having a variance σ_i^2. The weight of each observation will be denoted by w_i, where $w_i = 1/\sigma_i^2$. The model for the system has two parameters, *intercept* and *slope*:

$$y_i^{calc} = \text{intercept} + \text{slope } x_i$$

so the normal equations can be set up:

$$A_{11} \text{ intercept} + A_{12} \text{ slope} = b_1$$
$$A_{12} \text{ intercept} + A_{22} \text{ slope} = b_2$$
$$A_{11} = \Sigma w_i$$
$$A_{12} = \Sigma w_i x_i$$
$$A_{22} = \Sigma w_i x_i^2$$
$$b_1 = \Sigma w_i y_i^{obs}$$
$$b_2 = \Sigma w_i x_i y_i^{obs}$$
$$D = A_{11} A_{22} - A_{12}^2$$

The normal equations matrix can be readily inverted and the result is

$$P_{11} = A_{22}/D$$
$$P_{22} = A_{11}/D$$
$$P_{12} = -A_{12}/D$$

$$\rho = \frac{P_{12}}{P_{11} P_{22}} = \frac{-A_{12}}{A_{11} A_{22}}$$

where P_{11} is the variance on the intercept, P_{22} is the variance on the slope, P_{12} is the covariance between the slope and intercept and ρ is the correlation coefficient. Notice that these quantities depend only on the weights of the observations and the points x_i where the observations were made. Careful design of the experiment can control the magnitude of the covariance term in particular. To see this, let us suppose that the weights are all equal and that the observations are made at points symmetrically and equally spaced about the zero on the x axis. In that case, A_{12} is zero and consequently the covariance term P_{12} is also zero. By choosing to make the measurements at those points we have ensured that the parameters are completely uncorrelated! At the other extreme, suppose we are only able to make measurements in a small region remote from the origin on the x axis. In that case the correlation coefficient will be nearly equal to one. What this illustrates is the importance of experimental design, not only in determining the parameter errors by control of the experimental errors but also in determining the correlation coefficients, and hence the errors of other quantities, as we shall see below.

Thus far we have assumed that the variance–covariance matrix of the observations has been determined experimentally. In the real world this is often a far from simple task. An easier task is to estimate the elements relative to each other. This matrix $\mathbf{M}'$ is proportional to the true variance–covariance matrix $\mathbf{M}$:

$$\mathbf{M}' = \frac{1}{\sigma^2}\,\mathbf{M}$$

The proportionality constant σ^2 is known as the variance of an observation of unit weight. If we take the weight matrix to be the inverse of $\mathbf{M}'$ the errors on the parameters will be given by

$$\mathbf{P} = \sigma^2(\mathbf{J}^{\mathrm{T}}\mathbf{WJ})^{-1}$$

The quantity σ^2 can be estimated. It is shown in Appendix 5 that the expectation value of $\mathbf{r}^{\mathrm{T}}\mathbf{M}^{-1}\mathbf{r}$ is $m - n$. Therefore

$$E\{\mathbf{r}^{\mathrm{T}}(\mathbf{M}')^{-1}\mathbf{r}\} = (m - n)\sigma^2$$

and an estimate of σ^2 is given by $\mathbf{r}^{\mathrm{T}}\mathbf{Wr}/(m - n)$. In this case the variances on the parameters should be multiplied by the estimate of the variance of the observation of unit weight:

$$\sigma_i^2 = \frac{\mathbf{r}^{\mathrm{T}}\mathbf{Wr}}{m - n}\,P_{ii}$$

This procedure is most commonly applied when it can be assumed that there is no correlation and that all the variances are equal. Then $\mathbf{M}'$ is taken as the identity matrix and weights seem to disappear from the least-squares system. The penalty for not determining σ^2 experimentally is greater uncertainty in the estimates of parameter errors since the estimate σ^2 is itself subject to error. Also, different tests must be applied on statistical hypotheses concerning the parameters (Section 6.5).

3.6.2 ERRORS ON COMBINATIONS OF PARAMETERS

It often happens that we wish to derive quantities from the parameters and so are interested in the errors on these quantities. *In such circumstances correlation between parameters must be taken into account.*

For example, we might want to know the stepwise formation constants having determined the overall formation constants

$$K_{hj} = \beta_h/\beta_j$$

Applying the error propagation formula for a quotient (Section 2.2.3) we obtain

$$\left(\frac{\sigma_K}{K}\right)^2 = \left(\frac{\sigma_h}{\beta_h}\right)^2 + \left(\frac{\sigma_j}{\beta_j}\right)^2 - 2\frac{(COV)_{hj}}{\beta_h\beta_j}$$

The importance of the covariance term is that if it is negative the variance of the quotient will be increased, perhaps substantially.

Covariances between quantities derived from parameters may be calculated by linearizing the relationship, as in Section 2.2.2.

3.7 GEOMETRICAL INTERPRETATION OF THE LINEAR LEAST-SQUARES SYSTEM

The residuals are defined as

$$\mathbf{r} = \mathbf{y}^{obs} - \mathbf{Jp}$$

We will drop the superscript obs from y as there is no ambiguity. The sum of squares to be minimized is therefore

$$\begin{aligned}
S = \mathbf{r}^T\mathbf{Wr} &= (\mathbf{y} - \mathbf{Jp})^T\mathbf{W}(\mathbf{y} - \mathbf{Jp}) \\
&= \mathbf{y}^T\mathbf{Wy} - 2\mathbf{p}^T\mathbf{J}^T\mathbf{Wy} + \mathbf{p}^T\mathbf{J}^T\mathbf{WJp} \\
&= \mathbf{y}^T\mathbf{Wy} - 2\mathbf{p}^T\mathbf{b} + \mathbf{p}^T\mathbf{Ap}
\end{aligned}$$

This is a generalized quadratic equation in the parameters, as the highest power to which any parameter is raised is 2. We can therefore utilize some of the properties of quadratic equations to give a geometrical interpretation of the least-squares process. Most of this discussion will be limited to the case of two parameters, but it is easy to see how the discussion can be generalized.

When there are two parameters we can rewrite the quadratic as

$$\mathbf{p}^T\mathbf{Ap} - 2\mathbf{p}^T\mathbf{b} + \mathbf{y}^T\mathbf{Wy} - S = 0$$
$$A_{11}p_1^2 + 2A_{12}p_1p_2 + A_{22}p_2^2 - 2p_1b_1 - 2p_2b_2 + c = 0$$

This is of the same form as the equation for a general conic:

$$ax^2 + 2hxy + by^2 + 2gx + 2fy + c = 0$$

with the correspondences $p_1 = x$ and $p_2 = y$. The conic should be an ellipse. The condition for this is that $ab - h^2$ is positive, and this condition is met when the normal equations matrix $\mathbf{A}$ is positive definite (see Appendix 3). Let us write D for the discriminant $ab - h^2$, $D = A_{11}A_{22} - A_{12}^2$. Then the coordinates λ, μ of the centre of the ellipse are given by

$$\lambda = \frac{-bg + hf}{D}, \qquad \mu = \frac{hg - af}{D}$$

Substituting the appropriate values we obtain

$$\lambda = \frac{A_{22}}{D}\, b_1 + \frac{-A_{12}}{D}\, b_2, \qquad \mu = \frac{-A_{12}}{D}\, b_1 + \frac{A_{11}}{D}\, b_2$$

We see that these are precisely the values of p_1 and p_2 that minimize S, as in Section 3.5.1! Therefore if we draw contours on which the objective function S is constant, as a function of the two parameters, these contours are ellipses centred on the least-squares estimates of the parameters.

Some properties of the general conic are as follows. The angle, α, between the principal axes of the ellipse and the parameter axes is given by

$$\tan 2\alpha = 2h/(a - b)$$

The eccentricity, ε, of the ellipse is found by solving the equation

$$t^2 - (a + b)t + (ab - h^2) = 0$$

for the two roots A and B. Then

$$\varepsilon^2 = (B - A)/B$$

Table 3.1. Data used for Figures 3.1 to 3.3

	Set 1					Set 2				
y^{obs}	1.1	1.9	3.1	4.1	4.9	1.1	1.9	3.1	4.1	4.9
x	-2	-1	0	1	2	0	1	2	3	4
A	$\begin{bmatrix} 5 & 0 \\ 0 & 10 \end{bmatrix}$					$\begin{bmatrix} 5 & 10 \\ 10 & 30 \end{bmatrix}$				
b	$\begin{bmatrix} 15.1 \\ 9.8 \end{bmatrix}$					$\begin{bmatrix} 15.1 \\ 40.0 \end{bmatrix}$				
p^{min}	3.02 ± 0.047					1.06 ± 0.081				
q^{min}	0.98 ± 0.033					0.98 ± 0.033				
ρ	0					-0.8161				
α	0					-58°				
ε	0.7071					0.9775				

Let us investigate the effect of experimental design on the shape and size of the ellipses. For this purpose we shall examine a very simple model

$$y^{\text{calc}} = p + qx$$

in which we fit a straight line to the two parameters, p and q—slope and intercept. The 'observed' data consists of five synthetic values derived from the line $y = 3 + x$ by adding or subtracting an error of 0.1. We shall fit the data by the method of unit-weighted least squares, with two sets of the independent variable, as in Table 3.1. The two sets differ only in the shift of origin from 0,0 to 0, -2. The contours for $S = 1$, 5 and 10 are plotted for the two sets of data in Figures 3.1 and 3.2. Both contour maps are on the same scale and both parameter axes are drawn on the same scale. There are a number of points to

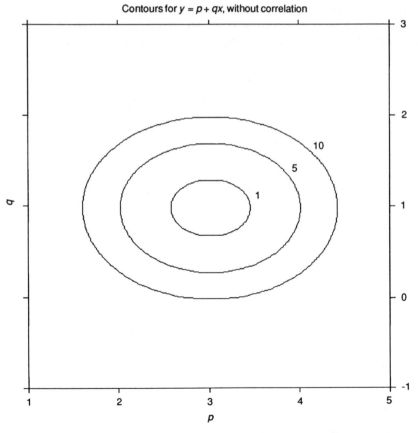

Figure 3.1. Contours of the sum of squared residuals for the model $y = p + qx$ with no correlation between p and q. The experimental data are given as set 1 in Table 3.1

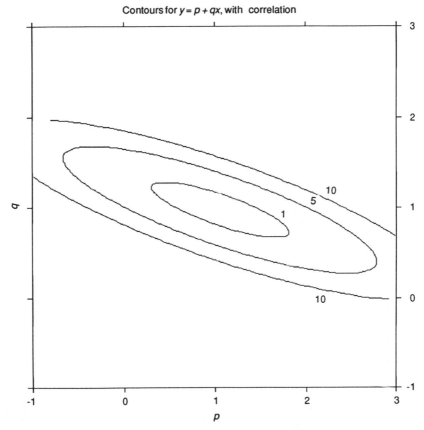

Figure 3.2. Contours of the sum of squared residuals for the model $y = p + qx$ with correlation between p and q. The experimental data is given as set 2 in Table 3.1

notice from the contour maps and the numerical results:

(a) The calculated value of the slope is the same in both cases. After allowing for the shift along the x axis, the value of the intercept is also the same. Thus the parameter values have not been materially changed by the shift of origin, which is comforting as the only difference between the two sets is the choice of origin.

(b) The standard deviation on the intercept is larger with set 2. This is intuitively reasonable, as in set 1 the value $x = 0$, at which the intercept is calculated, is in the middle of the data, whereas in set 2 it is at one side of the data. The larger standard deviation is reflected in the broader ellipse. This is shown more clearly in Figure 3.3, where the ellipses for $S = 5$ are plotted on the same centre. (The two x scales are respectively at the top and the bottom of the plot.)

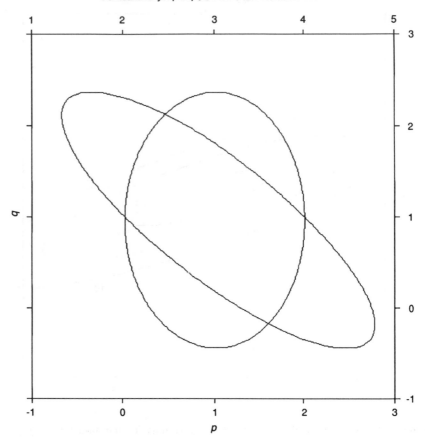

Figure 3.3. A comparison of contours for $S = 5$ for the model $y = p + qx$ with the two data sets in Table 3.1. The upper scale refers to the uncorrelated case (smaller ellipse)

(c) For set 1 the correlation coefficient is zero because the arrangement of the data values is symmetrical about $x = 0$. Therefore for set 1 the principal axes of the ellipse coincide with the parameter axes. For set 2 the correlation coefficient is negative. This means that there is a negative correlation between the errors on the slope and intercept. If the intercept were to be increased, the slope would have to be decreased. Also the principal axes of the ellipse are inclined with respect to the parameter axes, as a direct result of there being non-zero correlation.

(d) The eccentricity of the ellipse has increased from 0.7 for set 1 to 0.9775 for set 2. This is principally an effect of the change of origin. By following this through we see that the eccentricity becomes large as the difference

$A_{22} - A_{11}$ becomes large. In the limit $A_{11} \ll A_{22}$ the eccentricity becomes unity and the normal equations are singular. The main implication for this concerns the approximate location of the minimum for initial parameter estimation in the non-linear least-squares procedure (Section 4.2). If the ellipses in the region of the minimum are very elongated it will be difficult to obtain good parameter estimates.

In the general case for n parameters we can draw parameter ellipses by holding constant all but one pair of parameters. When there are three parameters the surfaces of constant S are parabolic ellipsoids. In other cases we would have to describe the surfaces as parabolic hyperellipsoids.

Note. A related ellipse can be derived by noting the relationships between A_{11}, A_{22} and A_{12} and the standard deviations and correlation coefficient of the parameters:

$$\sigma_p^2 = \frac{A_{22}}{D} \frac{S^{\min}}{m-n}$$

$$\sigma_q^2 = \frac{A_{11}}{D} \frac{S^{\min}}{m-n}$$

$$(COV)_{pq} = \frac{-A_{12}}{D}$$

$$\frac{\delta p^2}{\sigma_p^2} - \frac{2\rho_{pq}\, \delta p\, \delta q}{\sigma_p \sigma_q} + \frac{\delta q^2}{\sigma_q^2} = 1 - \rho_{pq}^2$$

This is known as the error ellipse. The differences between the parameters and their values at the minimum are δp and δq. The error ellipse, which has been discussed by a number of authors, allows one to see what the error on one parameter would be if the other had no error (the conditional standard deviation). It also facilitates the visualization of the joint probability of two parameters having given values. This joint probability is greater than 0.606 times the joint probability that the parameters have the least-squares values if the parameters lie inside the error ellipse. Clifford (1973) gives a clear description of the error ellipse.

3.8

3.8.1   THE METHOD OF ORTHOGONAL DECOMPOSITION

The essence of this method of minimizing an objective function is that it avoids the formation and solution of the normal equations by operating directly on the Jacobian. We shall first present the method, then show its equivalence to the normal equations method and finally critically compare the two methods.

The aim is to minimize the weighted sum of squared residuals, S. We will make the discussion simpler by subsuming the weights into the residuals, i.e. we will frame the discussion in terms of weighted residuals:

$$S = \mathbf{r}^T\mathbf{r} \leftarrow \mathbf{r}^T\mathbf{L}^T\mathbf{L}\mathbf{r} \leftarrow \mathbf{r}^T\mathbf{W}\mathbf{r}$$

A fundamental property of this function is that it is unchanged if the residual vector is left-multiplied by an orthogonal matrix. Let $\hat{\mathbf{r}} = \mathbf{H}^T\mathbf{r}$. As the inverse of an orthogonal matrix is its transpose, an orthogonal matrix multiplied by its transpose gives an identity matrix. Therefore

$$\mathbf{r} = \mathbf{H}\hat{\mathbf{r}} \quad \text{and} \quad S = \mathbf{r}^T\mathbf{r} = \hat{\mathbf{r}}^T\mathbf{H}^T\mathbf{H}\hat{\mathbf{r}} = \hat{\mathbf{r}}^T\hat{\mathbf{r}}$$

We now restate the problem by defining the residuals and subjecting $\mathbf{J}$ to an orthogonal decomposition, $\mathbf{J} = \mathbf{HR}$:

$$\Delta\mathbf{y} = \mathbf{y}^{obs} - \mathbf{y}^{initial}$$
$$\mathbf{r} = \Delta\mathbf{y} - J\,\Delta\mathbf{p} \quad (\mathbf{r} \leftarrow \mathbf{Lr},\ \Delta\mathbf{y} \leftarrow \mathbf{L}\,\Delta\mathbf{y},\ \mathbf{J} \leftarrow \mathbf{LJ};\ \mathbf{L}^T\mathbf{L} = \mathbf{W})$$
$$= \Delta\mathbf{y} - \mathbf{HR}\,\Delta\mathbf{p}$$

$\mathbf{H}$ is an orthogonal $m \times m$ matrix. The residual is then left-multiplied by $\mathbf{H}^T$:

$$\hat{\mathbf{r}} = \mathbf{c} - \mathbf{R}\,\Delta\mathbf{p}$$
$$\mathbf{c} = \mathbf{H}^T\,\Delta\mathbf{y}$$

The transforming matrix $\mathbf{H}$ is chosen so that $\mathbf{R}$ is zero everywhere except in the uppermost $n \times n$ triangle:

$$\mathbf{R} = \left[\begin{array}{cccc} R_{11} & R_{12} & \cdots & R_{1n} \\ 0 & R_{22} & \cdots & R_{2n} \\ \vdots & \cdots & \cdots & \cdots \\ 0 & 0 & 0 & R_{nn} \\ \hline & & \mathbf{0} & \\ \vdots & & & \end{array}\right] \begin{array}{l} \\ \\ n \text{ rows} \\ \\ \\ m - n \text{ rows} \end{array}$$

$$\mathbf{R} = \left(\frac{\mathbf{R}_{11}}{\mathbf{0}}\right)$$

We partition $\mathbf{c}$ in the same way into an upper portion, $\mathbf{c}_1$, of n elements and a lower portion, $\mathbf{c}_2$, of $m - n$ elements. The transformed residuals can then be written as

$$\hat{\mathbf{r}} = \left(\frac{\mathbf{c}_1 - \mathbf{R}_{11}\,\Delta\mathbf{p}}{\mathbf{c}_2 - \mathbf{0}\,\Delta\mathbf{p}}\right)$$

The sum of squares now becomes

$$S = ((\mathbf{c}_1 - \mathbf{R}_{11}\,\Delta\mathbf{p})^{\mathrm{T}} \mid \mathbf{c}_2^{\mathrm{T}})\left(\frac{\mathbf{c}_1 - \mathbf{R}_{11}\,\Delta_p}{\mathbf{c}_2}\right)$$

$$= (\mathbf{c}_1 - \mathbf{R}_{11}\,\Delta\mathbf{p})^{\mathrm{T}}(\mathbf{c}_1 - \mathbf{R}_{11}\,\Delta\mathbf{p}) + \mathbf{c}_2^{\mathrm{T}}\mathbf{c}_2$$

The sum of squares is minimized by setting the first component to zero:

$$\mathbf{R}_{11}\,\Delta\mathbf{p} = \mathbf{c}_1$$

and the minimal value is $\mathbf{c}_2^{\mathrm{T}}\mathbf{c}_2$.

The algorithm can be simply summarized. Firstly, orthogonally decompose the Jacobian into an upper triangular matrix, then left-multiply the $\Delta\mathbf{y}$ vector to obtain $\mathbf{c}$ and finally solve the n equations for the shifts. These equations are particularly easy to solve by the process of backward substitution because of the upper triangular nature of $\mathbf{R}$.

The algorithm is theoretically identical to the one in which the normal equations are formed and then solved. To demonstrate this we may write out the normal equations and perform the substitution $\mathbf{J} = \mathbf{HR}$:

$$(\mathbf{J}^{\mathrm{T}}\mathbf{J})\,\Delta\mathbf{p} = \mathbf{J}^{\mathrm{T}}\,\Delta\mathbf{y} \equiv \mathbf{R}^{\mathrm{T}}\mathbf{H}^{\mathrm{T}}\mathbf{HR}\,\Delta\mathbf{p} = \mathbf{R}^{\mathrm{T}}\mathbf{H}^{\mathrm{T}}\,\Delta\mathbf{y}$$

$$\equiv \mathbf{R}^{\mathrm{T}}\mathbf{R}\,\Delta\mathbf{p} = \mathbf{R}^{\mathrm{T}}\mathbf{H}^{\mathrm{T}}\,\Delta\mathbf{y}$$

$$\equiv \mathbf{R}\,\Delta\mathbf{p} = \mathbf{H}^{\mathrm{T}}\,\Delta\mathbf{y}$$

Note also that $\mathbf{J}^{\mathrm{T}}\mathbf{J} = \mathbf{R}^{\mathrm{T}}\mathbf{R}$. One may wonder, if the two methods are theoretically equivalent, why orthogonal decomposition was developed at all. There are two reasons. Firstly, they are not equivalent in practice because of the fact that numbers are computed with finite precision. Secondly, one particular form of orthogonal decomposition allows greater theoretical insight into the least-squares calculation This is the so-called singular value decomposition which we discuss below.

Orthogonal decomposition is supposed to be better at handling the situation in which the Jacobian is nearly singular. Lawson and Hanson (1974), who analysed the method in great detail, suggest that it is equivalent to working in double precision as compared to the normal equations method. However, it is far from clear whether this advantage is of use in practical computations using experimental data. In order to compare the two approaches I have incorporated both into a program used for many years to analyse infrared and Raman spectra by curve decomposition. The results have been compared for both synthetic and experimental data. The models are all non-linear, so it is necessary to protect the refinement against divergence by incorporating a Marquardt parameter (see Section 4.5.3). This was done in the orthogonal decomposition method by the process described by Lawson and Hanson; the only new feature was the method of calculating the cut-off value:

$$\lambda_c = 1/\max(1/R_{kk}) \qquad (k = 1 - n)$$

3.8.1.1 Storage

I used a procedure that requires the whole Jacobian to be stored, as opposed to a single row needed in the normal equations calculation. This can be obviated in the sequential method described by Lawson and Hanson, but at the expense of a somewhat slower refinement.

3.8.1.2 Speed

I found little difference in overall speed on the three sets of data processed. The largest data set contained 175 points and 21 parameters.

3.8.1.3 Convergence

Both methods converged to almost identical values, both in terms of the parameters and of the standard deviations. I was particularly interested to see whether orthogonal decompositions would give better results for the decomposition of two (Lorentzian) bands which overlap each other to a considerable extent. The question has the following experimental basis. We have observed that when two bands overlap each other extensively, the results of curve resolution tend not to be consistent across a series of related spectra. It was of interest to see if this inconsistency could be removed by using a method that has greater intrinsic numerical precision.

A spectrum was synthesized consisting of two Lorentzian bands with equal half-width, separated by one quarter of the half-width. Noise was added at the 1% level. The spectrum and resolved components are shown in Figure 3.4 and the numerical results in Table 3.2. These results are very interesting. They seem to show that the errors in the parameter values are not affected by the numerical precision of the calculations. Rather, these errors are due to the nature of the model and its sensitivity to errors in the data. It is reassuring to note, however, that the parameter values are within one standard deviation of the 'true' values.

In conclusion, therefore, it appears that the orthogonal decomposition method shows no great advantage or disadvantage compared to the normal equations method, even when the Jacobian is approaching singularity. (In the case above, two correlation coefficients are greater than 0.995 and four others are greater than 0.985.) In non-linear models precision is immaterial, since the error due to linearization is dominant, except at convergence.

3.8.2 SINGULAR VALUE DECOMPOSITION

In this variation we can suppose that the Jacobian is first decomposed to give an upper triangular $n \times n$ $\mathbf{R}_{11}$ and that this is then subjected to a diagonalization,

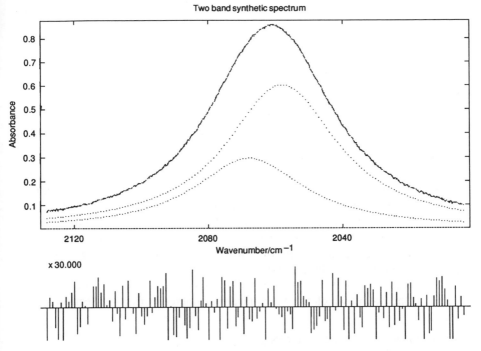

Figure 3.4. Upper part: data and model used for Table 3.2. Lower part: residuals $\times 30$

Table 3.2. A comparison of results of curve decomposition

		Normal equations	Orthogonal decomposition	'True value'
	Height	0.3327 ± 0.0290	0.3339 ± 0.0294	0.3
Band 1	Position	2067.40 ± 0.43	2067.37 ± 0.44	2068
	Half-width[a]	40.11 ± 0.30	40.12 ± 0.30	40
	Height	0.5665 ± 0.0291	0.5652 ± 0.0294	0.6
Band 2	Position	2057.76 ± 0.25	2057.75 ± 0.26	2058
	Half-width	39.98 ± 0.18	39.97 ± 0.19	40

[a] Full width at half-height.

again using orthogonal transformations. The result is the decomposition

$$\mathbf{J} = \mathbf{U S V}^{\mathrm{T}}$$

where $\mathbf{U}$ is an $m \times m$ orthogonal matrix, $\mathbf{V}$ is an $n \times n$ orthogonal matrix and $\mathbf{S}$ is a diagonal $n \times n$ matrix. After left-multiplying the residual by $\mathbf{U}^{\mathrm{T}}$ it becomes

$$\hat{\mathbf{r}} = \mathbf{U}^{\mathrm{T}} \Delta \mathbf{y} - \mathbf{S V} \Delta \mathbf{p}$$
$$= \mathbf{d} - \mathbf{S} \Delta \mathbf{q} \qquad (\mathbf{d} = \mathbf{U}^{\mathrm{T}} \Delta \mathbf{y}, \; \Delta \mathbf{q} = \mathbf{V} \Delta \mathbf{p})$$

The minimum in the sum of squared residuals is obtained by setting the first n components to zero:

$$\mathbf{d} = \mathbf{S}\ \Delta\mathbf{q}$$

Because of the diagonal nature of $\mathbf{S}$, the solution can be written down immediately:

$$\Delta q_k = d_k / S_{kk}$$
$$\Delta\mathbf{p} = \mathbf{V}^\mathrm{T}\ \Delta\mathbf{q}$$

Although this does not yield the parameter shifts directly, it is still very revealing. In particular, if any singular value tends towards zero, the corresponding parameters shift will tend to infinity. The singular values of the Jacobian should all be positive if the Jacobian has full column rank. In fact, the eigenvalues of the normal equations matrix $\mathbf{J}^\mathrm{T}\mathbf{J}$ are simply the squares of the singular values of $\mathbf{J}$, so a singular value of zero implies that the normal equations matrix is not positive definite. However singular value analysis, by concentrating on the Jacobian, avoids the complications inherent in dealing with the matrix product $\mathbf{J}^\mathrm{T}\mathbf{J}$. Singular value analysis is particularly helpful in understanding the way in which the Marquardt parameter works (see Section 4.5.3).

3.8.3   OTHER APPLICATIONS

Lawson and Hanson discuss a number of other applications of the orthogonal decomposition method. These include linear and non-linear constraints, and *under*-determined models. These are models in which the number of parameter exceeds the number of columns in the Jacobian (clearly such parameters cannot all be independent). A solution to such a model is given by the parameter set of minimum length, which is known as *least distance* programming. The solution involves *pseudo-inverse* matrices, which are easily obtained after orthogonal decompositions (see Appendix 2).

The 'total least-squares' method, which provides a way of allowing for errors in the independent variables, has been analysed in terms of a singular value decomposition by Golub and van Loan (1980).

4

Non-linear Least Squares

4.1 THE LINEAR APPROXIMATION TO A NON-LINEAR SYSTEM

We have so far assumed a linear relationship between the calculated quantities and the parameters, i.e. we have used linear models:

$$y_i^{\text{calc}} = \sum_{j=1,n} f_j(\mathbf{x}_i)\mathbf{p}_j$$

In this expression the coefficients of the parameters, $f_j(x_i)$, are independent of the parameters, and this defines the systems as linear ones. The quantities x_i are independent variables, usually assumed to be free from error.

There are many cases in which the functional relationship cannot be written in linear form. In that case we have to write

$$y_i^{\text{calc}} = f(\mathbf{p}, \mathbf{x}_i)$$

which simply means that the calculated quantity is a function for all the parameters p_j ($j = 1, n$) and all the independent variables x_i on which the ith calculated value depends. For example we could have an exponential relationship $y^{\text{calc}} = p\,e^{qx}$ where p and q are the parameters and x is the independent variable: the relationship is non-linear because the partial derivatives $\partial y/\partial p$ and $\partial y/\partial q$ both contain the parameters themselves. This kind of problem can be treated by the method of least squares by using a technique for successive approximations. The key to using this technique is to expand the function as a Taylor series about some value of the parameter set, $\mathbf{p}^{\text{initial}}$, say:

$$y_i^{\text{calc}} = y_i^{\text{initial}} + \sum_{j=1,n} (\partial f_i/\partial p_j^{\text{initial}})\,\Delta p_j$$

In writing this expression we are making two approximations, namely that the other terms in the Taylor series expansion can be ignored and that the infinitesimal dp_j can be replaced by the finite increment Δp_j. Both of these approximations are usually very good in the limit when Δp_j becomes small, i.e. when y_i^{calc} is nearly equal to y_i^{initial}.

With this approximation for the values of y_i^{calc} we can follow the same logic as was used in the case for linear least squares in order to minimize the objective function $\mathbf{r}^T\mathbf{Wr}$, by identifying the residuals as

$$r_i = y_i^{\text{obs}} - y_i^{\text{calc}}$$

$$= y_i^{\text{obs}} - y_i^{\text{initial}} - \sum_{j=1,n} (\partial f_i/\partial p_j^{\text{initial}})\, \Delta p_j$$

$$\mathbf{r} = \Delta\mathbf{y} \qquad - \qquad \mathbf{J}\,\Delta\mathbf{p}$$

In the linear case we have $\mathbf{r} = \mathbf{y}^{\text{obs}} - \mathbf{Jp}$, and the solution to the minimization is given by the normal equations $\mathbf{J}^T\mathbf{WJp} = \mathbf{J}^T\mathbf{Wy}^{\text{obs}}$. In the non-linear case the normal equations are

$$\mathbf{J}^T\mathbf{WJ}\,\Delta\mathbf{p} = \mathbf{J}^T\mathbf{W}\,\Delta\mathbf{y}$$

At first sight these equations look very similar to the ones obtained for a linear system, with $\mathbf{p}$ replaced by $\Delta\mathbf{p}$ and $\mathbf{y}$ replaced by $\Delta\mathbf{y}$. There are, however, some very significant complications in the apparently simple substitutions.

Firstly, there is an absolute need to estimate some initial parameter values, whereas in a linear model parameter estimation is optional. The problem of estimating values for the parameters is in general not a trivial one, and in some cases can become a major difficulty.

Secondly, the correction $\Delta\mathbf{p}$, when added to the initial parameter values, does not usually yield the correct least-squares estimates of the parameters immediately. Rather, the process must be repeated with $\mathbf{p} \leftarrow \mathbf{p} + \Delta\mathbf{p}$ as the new initial estimate of the parameters. Thus, the calculation becomes *iterative*, with the parameters being refined by successive approximation. This is a result of the inadequacy of the first-order Taylor series approximation.

Thirdly, the iterative nature of the process introduces other problems. Ideally the objective function S should decrease smoothly towards its minimum value as the iterations proceed, but it often happens that S may increase at some stage. This is known as *divergence*, and is also a result of the inadequacy of the first-order Taylor series approximation.

Finally, there is the problem of deciding when the iterations should be terminated. This is known as setting *convergence criteria*, and it is far from easy in the general case.

The non-linear least-squares procedure can be summarized as having two distinct phases: initial parameter estimation and then iterative refinement, guarding against divergence and checking at the end of each refinement cycle for final convergence. Many people try to program a non-linear system without protecting against divergence, and at first this may be satisfactory, but sooner or later divergence will occur. It is therefore preferable, when writing the program, to include protection against divergence in the planning stage because otherwise the program may need extensive rewriting.

At the conclusion of a non-linear refinement one obtains the usual information on parameter errors and correlation coefficients. However, it needs to be borne in mind that these values are in a statistical sense *biased* and should accordingly be treated with caution. The bias arises because the treatment of errors assumed a linear relationship between parameters and calculated values. With the non-linear relationship errors are obtained by means of a first-order Taylor series expansion, and this introduces the bias through neglect of the higher-order terms in the expansion.

4.2 ESTIMATION OF PARAMETERS

In all non-linear least-squares problems there is a need to furnish initial estimates of the parameters with which to begin the refinement. In some cases the parameter estimates need to be really close to the final values or the refinement will not work at all; in other cases the requirements are less stringent. It is difficult in general to predict what the requirement will be, so it is prudent in all cases to take some trouble to estimate the parameters as well as is practicable. Time taken to do this is often amply repaid in terms of trouble avoided. The first line to follow is to use all that is known about the physics and chemistry of the system under study, and of closely related systems. This will dictate the choice of model (Section 5.1), any constraints that may apply (Section 5.2) and which parameters are to be refined.

Ideally one would construct an expert system that could provide reliable parameter estimates by extrapolation or interpolation from known facts. Unfortunately all too often the known facts are too sparse to permit this and one is left with little more than intelligent guesswork. Parameter guesses will only be usable in simple systems. In more complicated systems one will need to resort to numerical techniques or simulation.

Numerical techniques have certain features in common. They all involve searching through the parameter space and they all require the application of some criteria to decide when satisfactory estimates have been found. The most commonly used criterion is that the minimum in the objective function should at least be bracketed.

One might imagine that a simple grid search would be satisfactory: each parameter varies in a number of discrete steps between some chosen minimum and maximum values and the objective function is calculated for each combination of values for different parameters. However, the grid search is rarely of any

Non-linear Least Squares

real use, for a number of reasons. Firstly, the number of times the objective function has to be calculated is equal to the number of steps raised to the power of the number of parameters. If each of six parameters has to be tried with ten trial values, there are 10^6 or one million combinations to be tried! Then there is a problem in deciding the limits and size of the grid mesh—the distance between grid points. With too large a mesh the minimum may be missed, but a smaller mesh requires many more calculations. Moreover, the best mesh size for one variable may be quite different from the best mesh size for another.

These points can best be understood with the aid of a graphical example. In Figure 4.1 we show the contours of the sum of squares for the function $y = p\,e^{-qx}$ with two parameters p and q. The 'observed' data were constructed

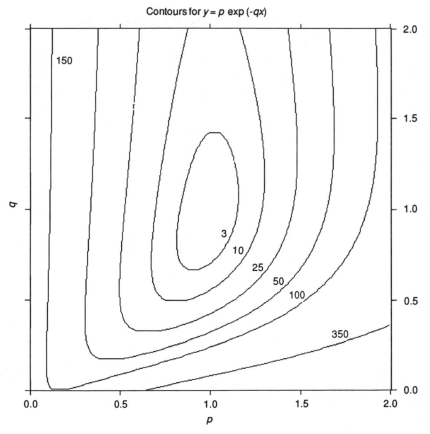

Figure 4.1. Contours of the sum of squared residuals (multiplied by 100) for the model $y = p\,e^{-qx}$

with $p = 1$, $q = 1$ and $x = 0, 10$; the error term was a random number taken
from a uniform distribution of mean zero and variance 0.1. The contour map
was constructed by allowing p and q to vary between 0 and 2, and calculating
those values of p and q that give a set of equal values of the objective function
S. As we have seen in Section 3.7, the contours should be elliptical in the
region of the minimum. Away from the minimum the contour map shows two
distinct regions: in the lower half the contour lines are nearly horizontal,
making it difficult to see where the minimum with respect to p lies, whilst in
the upper half the contour lines are almost vertical, making it difficult to locate
the region where q has its optimum value.

If we had no prior knowledge of the approximate location of the minimum,
how would we choose the region to search? A range of search values such as
$p, q = 0, 10, 20, \ldots$ would result in complete failure to bracket the minimum.
Should one then choose a larger or smaller grid? It is clear that in this situation
a good choice of grid would be fortuitous.

The situation is even more complicated when there are multiple minima.
Figure 4.2 shows part of the contour map for the function

$$y = p \, e^{-q^2 x}$$

with the two parameters p and q; the data were constructed in the same way
as for the exponential function above, but now there are minima at both
$q = -1$ and $q = +1$. The two halves of the map are mirror images of each
other about the line $q = 0$. When two such minima exist, the grid search must
be limited so as to encompass only one of them. The choice of minimum is
often determined by the physics of the system, as, for example, when all
parameters correspond to physical quantities that must have positive
numerical values. Another instance of multiple minima arises when
parameters are interchangeable, as in the sum of exponentials $e^{-px} + e^{-qx}$,
which has minima at $p = p_m$ and $q = q_m$, or alternatively at $p = q_m$ and $q = p_m$
(see Section 4.4). In this case the grid search should be constrained such that
p is always greater than q, or vice versa.

The grid search is an example of a so-called direct search method, i.e.
a method in which the parameters are varied directly rather than being
subjected to shifts calculated by the least-squares method. Various techniques
exist to improve the efficiency of direct search methods, but they all have
limitations. In practice they amount to methods of minimization that do not
depend on derivatives. The methods are described in detail in Box, Davies
and Swann (1969) and in Section 4.8. However, any full-blown minimization method suffers from the inherent defect that the convergence criteria
are all but impossible to set. If set too loosely the minimum will be
bracketed too broadly: if set too tightly, the calculation becomes very
lengthy.

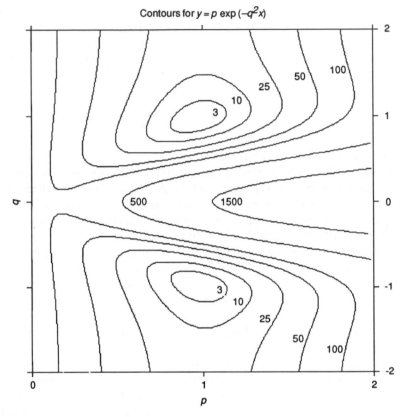

Figure 4.2. Contours of the sum of squared residuals (multiplied by 100) for the model
$y = p\mathrm{e}^{-q^2 x}$

4.3 SIMULATION

Perhaps the best way to estimate parameters is not a purely numerical one, but one that uses a combination of numerical calculation and human judgement— simulation. Of course, this is open to the objection that it introduces subjectivity, but in this instance subjectivity is actually desirable as, on the one hand, parameter estimates must appear to be meaningful and, on the other, the estimates are only to be used in the objective process of least-squares minimization.

The process of simulation consists of constructing an algorithm that obeys the model chosen for the system, and allows for the parameters to be varied individually. The object is to try to adjust the parameters in such a way as to bring y_i^{calc} close to y_i^{obs}. However, rather than use the least-squares criterion for the closeness of fit, the curves representing y^{calc} and y^{obs} should be plotted together, and the parameters adjusted to make them nearly superimposable. When they are close to each other, limitations of computer screen resolution make it difficult to see what is going on. It is then useful to plot the residuals $y_i^{obs} - y_i^{calc}$, magnified as appropriate, and to base the parameter adjustment on the residuals.

Usually each parameter affects the residuals in a different way—one parameter may affect them only in one region of the independent variable x, whilst another may affect them systematically over a range of x values. This is where human judgement becomes very useful as the human expert has an intuitive understanding of the relationship of the parameters to the experimentally observed quantities, built up with experience.

An example of a simulation display is shown in Figure 4.3. This illustrates an attempt to fit an infrared spectrum with two bands of Lorentzian shape. The observed data are plotted as a solid line and the calculated bands as dotted lines. In fact the original program that produced this plot shows all the curves on the screen in different colours to enable them to be more easily distinguished. Below the data are shown the residuals, magnified by a factor

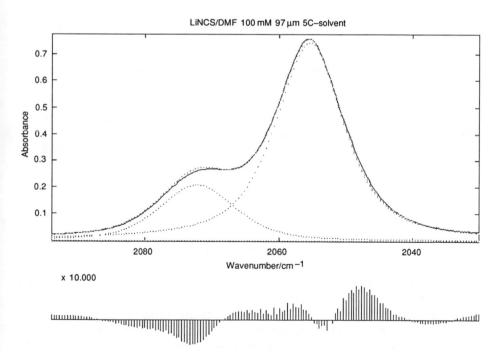

Figure 4.3. An example of a simulation display. Upper part: observed data (solid line), components and sum of components (dotted lines). Lower part: residuals $\times 10$

of 10. The residuals are shown as vertical bars. The model used for this simulation involves six parameters:

$$y_i^{\text{calc}} = \frac{h_1}{1 + [2(p_1 - v_i)/w_1]^2} + \frac{h_2}{1 + [2(p_2 - v_i)/w_2]^2} + \text{baseline}_i$$

where h_1, p_1 and w_1 are the height, position and width of band 1 and h_2, p_2 and w_2 are the height, position and width of band 2; v_i is the wavenumber at the ith point. The baseline was kept at zero. The left-hand minimum in the residuals coincides with the left-hand peak in the spectrum, so clearly the band height needs to be decreased. However, the right-hand maximum in the residuals does not coincide with an absorption maximum; this implies that it is the width of the right-hand band that needs to be changed.

Changing the parameters can be made into an interactive process by using a cursor, controlled by a light-pen, a mouse or cursor keys. The cursor is moved around the screen and its position is made to correspond to a parameter value. For instance, the height of the cursor, h_c, can be made to represent the height of a band, h_j: if the screen is set up such that the displayed cursor range is a, b in graphical units and the displayed parameter range is c, d in user units, the correspondence is given by

$$\frac{h_c - a}{b - a} = \frac{h_j - c}{d - c}$$

so that h_c may be calculated from h_j and vice versa. Likewise the horizontal position of the cursor can be made to correspond to a band position. The width parameter w_1 can be represented as a pair of vertical lines drawn on the screen at positions corresponding to $v = p_1 \pm w_1/2$, whose separation can be increased or decreased in correspondence with changes in w_1.

To make the simulation complete we need a way of selecting which band to work on. We may also wish to add a new band, or take one away, and we need to be able to use band shapes other than Lorentzian ones. Separate keys, such as function keys, can be assigned to these functions. One may wish to vary a baseline, which may be a polynomial, such as

$$\text{Baseline}_i = h_b + s\,\frac{v_i - v_0}{v_m - v_0} + q\left(\frac{v_i - v_0}{v_m - v_0}\right)^2 + \cdots$$

with parameters $h_b, s, q, \ldots$. Alternatively, it may be an exponential or some other form that depends on other parameters.

If the calculation of y_i^{calc} ($i = 1, m$) is sufficiently rapid the new simulation can be shown every time a parameter is changed. It can be speeded up by calculating only selected points, at the expense of loss of detail. It is convenient to have a facility for making large, medium or small adjustments to the parameters.

The effects of the variation of parameters in terms of the individual bands, the calculated spectrum and the residuals will be seen quickly. No numerical data are required, except for the initial values of the parameters which may be set up automatically with arbitrary values. With complex models it is useful to have a facility for numerical adjustment in order to be able to enter parameter estimates that are known to be reliable and to be able to store and recall complete parameter sets.

The objective of parameter adjustment in simulation is to try to reduce *systematic* trends in the residuals. It is therefore complementary to the least-squares minimization which reduces the *random* variations in the residuals. The advantage of simulation over purely numerical methods for estimating parameters is that it ensures that the trends in calculated values follow the trends in observed values. It is therefore useful to check that the model is adequate. For example, if a parameter is missing from the model that should be there, no amount of fiddling with the other parameters will remove those systematic trends in the residuals that are the result of the parameter being missing. When this parameter is added to the model the situation will be improved. It follows that the simulation should allow not only parameters to be varied but also that the model be varied. Variation of the model includes the elimination of one or more parameters. In this case one observes whether the elimination of a parameter has a significant effect upon the residuals.

Once a simulation program has been written it should be integrated with the least-squares refinement, so that when the parameters have been estimated they can be used directly in the refinement. The process may also be useful in the reverse direction, for building up complicated models: first parameters are estimated for a simplified model and then refined; new parameters are added to the model and are adjusted; the new model is refined and so forth. This is particularly useful for parameters that have only a small effect on the observations.

4.4 MULTIPLE MINIMA

When we minimize the weighted sum of squared residuals, $S = \Sigma \hat{r}^2$, with respect to the parameters, the minima are specified by setting the partial derivatives of the objective function with respect to each parameter equal to zero. The weights are subsumed into the residuals $\hat{r}$; for convenience we shall discuss in detail only the unit-weighted case, $S = \Sigma r^2$, $r_i = y_i^{obs} - y_i^{calc}$. The minima are given by

$$\frac{\partial S}{\partial p_k} = -2 \sum_{i=1,m} r_i \frac{\partial y_i^{calc}}{\partial p_k} = 0 \qquad (k = 1, n)$$

When we write $\partial y_i^{calc}/\partial p_k$ as J_{ik} and remove the factor -2, we obtain a set of n equations which we may call the gradient equations:

$$\sum_i J_{ik} y_i^{obs} = \sum_i J_{ik} y_i^{calc} \quad (k = 1, n); \qquad \mathbf{J}^T \mathbf{y}^{obs} = \mathbf{J}^T \mathbf{y}^{calc}$$

If the system is linear, $\mathbf{y}^{calc} = \mathbf{Jp}$, and the normal equations $\mathbf{J}^T \mathbf{y}^{obs} = \mathbf{J}^T \mathbf{Jp}$ result; as $\mathbf{J}$ is not a function of the parameters these equations have a unique solution and there is only one minimum. However, in the non-linear case the elements of $\mathbf{J}$ are functions of the parameters, and so the possibility of multiple minima exists. Writing the gradient equations as

$$\mathbf{J}^T (\mathbf{y}^{obs} - \mathbf{y}^{calc}) = 0$$

we have n homogeneous equations in the n parameters. We can immediately see two possibilities: either $\mathbf{J}$ has reduced rank or $\mathbf{J}^T \mathbf{y}^{obs} = \mathbf{J}^T \mathbf{y}^{calc}$. The possibility of the Jacobian not having full column rank arises because it is a function of the parameters.

Three simple cases in which multiple minima occur can be easily identified—models that are at least quadratic in one parameter, models in which parameters can be interchanged and models involving trigonometric functions such as sin and cos.

Let us take an artificially simple model of the first case, which is quadratic in a single parameter:

$$y_i^{calc} = (a + bp + cp^2) x_i$$

This model fits a straight line through the origin to a function of the one parameter p and some constants a, b and c. The gradient equation is

$$\sum_i (b + 2cp) x_i y_i^{obs} = \sum_i (b + 2cp)(a + bp + cp^2) x_i^2$$

which simplifies to

$$(b + 2cp) \sum_i x_i y_i^{obs} = (b + 2cp)(a + bp + cp^2) \sum_i x_i^2$$

which has one solution at $b + 2cp = 0$:

$$p_1 = -b/2c$$

In this solution the rank of the Jacobian is zero. The other solutions are given by

$$p_2 = p_m \qquad \text{and} \qquad p_3 = -p_m - b/c$$

where

$$p_m = \frac{-b + \sqrt{(b^2 - 4cz)}}{2c}, \qquad z = a - \frac{\sum_i x_i y_i^{obs}}{\sum_i x_i^2}$$

It is easy to show that p_2 and p_3 give identical values of the objective function S. Hence they constitute the two minima and p_1 constitutes the maximum, which occurs at the average value of p_2 and p_3. A plot of the sum of squared residuals against the parameter p is shown in Figure 4.4. The 'observed' data for this example were constructed with the constants $a = 1$, $b = -10$ and $c = 1$, and no added noise. The true minimum is at $p = 2$, the false minimum is at $p = 8$ and the maximum is at $p = 5$.

It is interesting to note that the Jacobian, $(b + 2cp)x_i$, is a zero vector at the maximum. This is a particular instance of the Jacobian having reduced rank. The region in parameter space where this occurs is between two minima. In general there will be a region between two minima where both the Jacobian and the normal equations matrix have rank less than n. Here is an informal proof. In the least-squares refinement the shift vector points 'downhill'

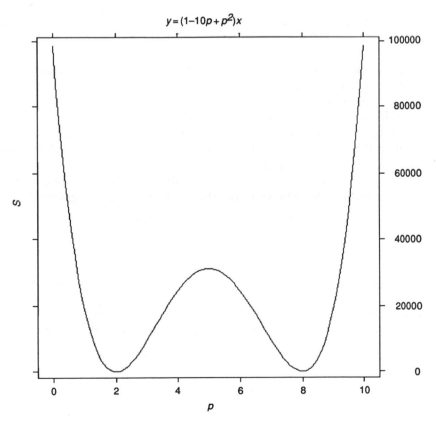

Figure 4.4. The sum of squared residuals for the model $y = (1 - 10p + p^2)x$ as a function of the parameter p

towards a minimum as long as the normal equations matrix is positive definite (see Section 4.5). When there are two minima there exists a region where the shift vector cannot simultaneously point downhill towards both minima. In this region the normal equations matrix cannot be positive definite, so it must be singular. It is clear that the region includes the point at which the objective function has a maximum.

In this example there is nothing to distinguish between the true minimum and the false one, at least as far as the value of the objective function is concerned. A more pernicious, if again rather artificial, example can be constructed with the model

$$y_i^{calc} = (1 - 3p + p^3)x_i$$

The gradient equation for this model is

$$(-3 + 3p^2) \sum_i x_i y_i^{obs} = (-3 + 3p^2)(1 - 3p + p^3) \sum_i x_i^2$$

If we take the true parameter value to be $p = -3$, this clearly corresponds to the true minimum. Other extrema are at $p = -1$ and $p = 1$ $(3p^2 - 3 = 0)$.[*] The plot of the objective function is shown in Figure 4.5. It can be seen that when $p = 1$ there is a false minimum where the objective function has a higher value than at the true minimum. Moreover, the Jacobian is a zero vector at the false minimum, so the normal equation matrix is also singular there!

It is difficult to generalize from these primitive examples; they serve to illustrate the kinds of problems that may arise with non-linear models. What we can state is that whenever a product such as $p^a q^b$ occurs in a model there will be at least $a + b$ minima in the objective function.

Common examples of models that are quadratic functions of one or more parameters include the following line shapes used to analyse spectra:

$$y_i^{calc} = h\, e^{-\ln 2[4(p - x_i)^2/w^2]} \qquad \text{(Gaussian)}$$

$$y_i^{calc} = \frac{h}{1 + 4(p - x_i)^2/w^2} \qquad \text{(Lorentzian)}$$

where the parameters h, p and w are the height, position and full width at half-height of the function. Figure 4.2 shows a contour map for a simplified Gaussian function. These models will have minima at $w = \pm w_m$. In this example we can eliminate one minimum by a change of variable, $z = w^2$:

$$y_i^{calc} = h\, e^{-\ln 2[4(p - x_i)^2/z]} \qquad \text{(Gaussian)}$$

$$y_i^{calc} = \frac{h}{1 + 4(p - x_i)^2/z} \qquad \text{(Lorentzian)}$$

---

[*] The gradient equation is quintic in the parameter in this example. It has three real solutions as given; the other two solutions are complex.

$y = (1-3p+p^3)x$

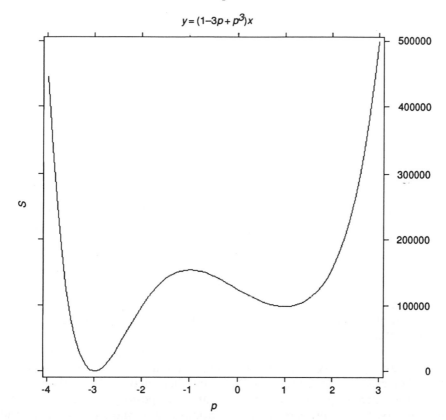

Figure 4.5. The sum of squared residuals for the model $y = (1 - 3p + p^3)x$ as a function of the parameter p

Note that in these examples there are two values of the width parameter that give the same value for y^{calc}, giving rise to two equal minima in the objective function. However, the position parameter, whilst it appears as a quadratic term in the model, is unique as there is a unique position of the maximum in y^{calc}.

Models in which parameters can be interchanged include all those that consist of a sum of similar functions, such as a sum of exponentials, Gaussians or Lorentzians. To illustrate this consider the simplest of these sums, that of two normalized exponentials

$$y_i = e^{px_i} + e^{qx_i}$$

where p and q are the two parameters of this model. If one minimum exists at $p = p_m$ and $q = q_m$ another minimum exists at $p = q_m$ and $q = p_m$. Moreover, it is obvious that S has the same value at the two minima. It may seem that this is a trivial case, since it only affects the ordering of the parameters, but there is one important corollary. Whenever the two parameters p and q are equal, both columns of the Jacobian are identical and the normal equations matrix $\mathbf{J}^T\mathbf{WJ}$ is singular. In general, whenever two functionally similar quantities occurring in a sum have the same parameter values, the Jacobian will have reduced rank, and there will be a region where the normal equations matrix is singular. If, in the process of iterative refinement, the parameter values fall on this region, the refinement will fail. Obviously the risk of this happening increases as parameters such as p_m and q_m become closer to each other.

Figure 4.6 shows a contour map for the sum of two exponentials. The line of the singular normal equations matrix is defined by $p = q$. The 'observed' data, shown in Table 4.1, were constructed with $p = 0.5$ and $q = 1.5$, and added noise taken from a uniform distribution of mean zero and variance 0.1.

The trigonometric case is straightforward. Suppose the model is $y_i^{\text{calc}} = \sin(px_i)$. The gradient equation is then

$$\sum_i x_i \cos(px_i) y_i^{\text{obs}} = \sum_i x_i \cos(px_i)\sin(px_i)$$

If p_m is a solution to this equation, then so also are $p_m + 2n\pi$ $(n = 1, 2, 3, \ldots)$. The multiple minima are eliminated by constraining the parameter to values between 0 and 2π.

These examples illustrate how easily multiple minima can occur in a non-linear least-squares system. The number of minima can increase alarmingly fast. For example, in fitting two Gaussians there may be four minima, two each for interchange of the functions and for positive or negative width parameters. With some care similar situations can be spotted in advance and suitable precautions taken. This means reformulating the model by, for example, changing the designated parameters. However, the general case is one in which the functional form of the model is perhaps too complex to admit of this simple analysis, as, for example, when y^{calc} is an implicit function of the parameters. The question then arises: what is the risk of the refinement leading to a false minimum? The existence of false minima can be verified by starting the refinement from different initial parameter values. Only if there is consistent convergence to two different minima will it be necessary to make a choice between two possible models. Perhaps the most common reason for false minima appearing has, in the past, been premature convergence, resulting from the use of incorrect convergence criteria.

If multiple minima cannot be eliminated from the model by reformulation it will be necessary to choose the initial values for the parameters rather carefully. These initial values will ideally be close to the required minimum and distant from the minimum, which is not required. This is where simulation is invaluable, since it enables the user to start with plausible parameter values.

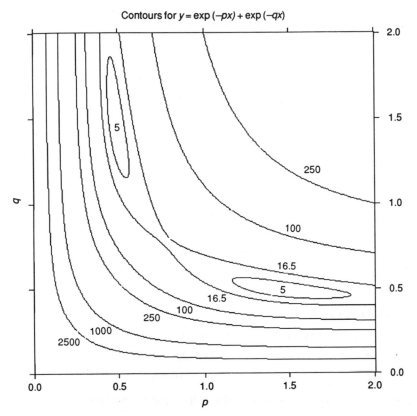

Figure 4.6. Contours of the sum of squared residuals (multiplied by 100) for the model $y = e^{px} + e^{qx}$. The observed data are in Table 4.1

Table 4.1. Data used for the double exponential model
$$y_i = e^{px_i} + e^{qx_i}$$

x	y^{obs}
0	1.9837
1	0.8393
2	0.4305
3	0.2441
4	0.1248
5	0.0981
6	0.0549
7	0.0174
8	0.0249
9	0.0154
10	0.0127

4.5

4.5.1   NON-LINEAR REFINEMENT

As we have seen in Section 4.1, the calculation of the parameters of a non-linear model is an iterative process. Let us consider the first iteration. We begin by assigning initial values to the parameters, $\mathbf{p}^{\text{initial}}$, and from these calculate the Jacobian, $\mathbf{J}^{\text{initial}}$, $\mathbf{y}^{\text{initial}}$ and the value of the objective function, $\mathbf{S}^{\text{initial}}$. The normal equations are set up and solved for the parameter shift vector, $\Delta\mathbf{p}$ as follows. The model is linearized by means of a first-order Taylor series expansion:

$$y_i^{\text{calc}} = y_i^{\text{initial}} + \sum_{j=1,n} (\partial y_i/\partial p_j^{\text{initial}})\, \Delta p_j$$

$$\mathbf{y}^{\text{calc}} = \mathbf{y}^{\text{initial}} + \mathbf{J}\,\Delta\mathbf{p}; \qquad \mathbf{J} = \mathbf{J}^{\text{initial}}$$

Since $\mathbf{r} = \mathbf{y}^{\text{obs}} - \mathbf{y}^{\text{calc}}$, the minimum of $\mathbf{r}^{\text{T}}\mathbf{W}\mathbf{r}$ is given by

$$\frac{\partial S}{\partial \mathbf{p}} = -2\mathbf{J}^{\text{T}}\mathbf{W}\mathbf{r} = 0$$

and the normal equations become

$$\mathbf{J}^{\text{T}}\mathbf{W}(\mathbf{y}^{\text{obs}} - \mathbf{y}^{\text{initial}} - \mathbf{J}\,\Delta\mathbf{p}) = 0$$
$$\mathbf{J}^{\text{T}}\mathbf{W}\mathbf{J}\,\Delta\mathbf{p} = \mathbf{J}^{\text{T}}\mathbf{W}\,\Delta\mathbf{y}$$
$$\Delta\mathbf{y} = \mathbf{y}^{\text{obs}} - \mathbf{y}^{\text{initial}}$$

The shift vector is then added to the initial parameter vector:

$$\mathbf{p}^{\text{initial}} \leftarrow \mathbf{p}^{\text{initial}} + \Delta\mathbf{p}$$

and values for $\mathbf{y}^{\text{calc}}$ and S are obtained from the new 'initial' set of parameters.

If the model were linear the new value of S would be the minimum value and the new parameters would lie at the centre of the ellipsoid of constant S with respect to the parameters, which passes through the initial point, $\mathbf{p}^{\text{initial}}$. In a non-linear model, the linear approximation means that the centre of the approximated ellipsoid through the initial point does not coincide with the position where the objective function has been minimized. The actual contours are not ellipsoidal. We could say that the shift vector is wrong in both direction and magnitude. Assuming for the moment that S is less than S^{initial}, we must designate the new parameter set as $\mathbf{p}^{\text{initial}}$ and repeat the whole refinement cycle.

As the refinement proceeds the first-order Taylor series expansion of y^{calc} becomes a good approximation. The actual contours of the objective function become ellipsoidal, and at some stage the objective function is as near to its minimum value as possible. We say that at that stage the refinement has *converged*.

However, it may happen that the calculated shift vector leads to a point in parameter space where the objective function has a *higher* value than the point of origin. This phenomenon is known as *divergence*. Divergence is a problem common to all non-linear systems, and will usually occur when the initial parameter estimates are not very good. What 'very good' means varies from one system to another. It is therefore important that the refinement process should be protected against divergence. Various forms of protection have been devised, as we shall now discuss.

4.5.2 SHIFT-CUTTING

In this method it is assumed only that the *length* of the shift vector is seriously in error, whilst it points in approximately the right direction. The length of the shift vector can be reduced so as to lead to a point where the sum of squares is lower than the starting point. This is a reasonable assumption for, as long as the normal equations matrix $\mathbf{J}^T\mathbf{WJ}$ is positive definite, the shift vector will point 'downhill', as we will now show.

The gradient of the objective function with respect to the parameters is $-2\mathbf{J}^T\mathbf{W}\,\Delta\mathbf{y}$ at the point where the parameter vector is $\mathbf{p}^{\text{initial}}$. Therefore the vector of steepest descent, which is 'perpendicular' (orthogonal) to the gradient, is $2\mathbf{J}^T\mathbf{W}\,\Delta\mathbf{y}$. The dot product of any two vectors is positive if the angle between them lies in the range $-90° < \theta < 90°$. The dot product between the shift and steepest descent vectors is

$$2\,\Delta\mathbf{y}^T\mathbf{WJ}(\mathbf{J}^T\mathbf{WJ})^{-1}\mathbf{J}^T\mathbf{W}\,\Delta\mathbf{y}$$

The matrix $\mathbf{WJ}(\mathbf{J}^T\mathbf{WJ})^{-1}\mathbf{J}^T\mathbf{W}$ is known as the hat matrix and has the property (see Section 6.4) that if the rank of the Jacobian is n it has exactly n unit eigenvalues. This being so, the quadratic form above must be positive and the angle between the shift vector and the vector of steepest descent will be less than $90°$; the shift vector must point in a direction along which the objective function at first decreases.

Shift-cutting usually begins by halving the length of the shift vector. If the sum of squares is convergent the shortened shift vector is accepted; otherwise it is halved again, and so on. A sophisticated version of the step-cutting process optimizes the length of the reduced shift vector, by some process of one-dimensional minimization.

Shift-cutting is a useful and simple method of protecting against divergence. It has two main disadvantages. Firstly, if the length of the shift vector is much too long it requires a substantial number of calculations of S, especially if the length is optimized. More seriously, if the direction of the shift vector is bad, shift-cutting will only lead to a small improvement in S, and the overall number of iterations needed to attain convergence will become large.

4.5.3 ROTATION (THE MARQUARDT PARAMETER)

In order to improve the direction of the shift vector it needs to be rotated so that it points towards the minimum. Although this cannot be achieved, rotation towards the direction of steepest descent is possible. The direction of steepest descent is perpendicular to the contour lines of S, so S is guaranteed to decrease rapidly in this direction. The steepest descent direction is not itself a good one for least-squares minimization, because the minimum is not usually in that direction. Indeed, the local direction of steepest descent may point away from the direction of the minimum. However, a small rotation of the shift vector towards the direction of steepest descent has proved to be a very effective strategy.

A way of rotating the shift vector towards the direction of steepest descent was proposed independently on at least five occasions by Levenberg (1944), Girard (1958), Wynne (1959), Morrison (1960) and Marquardt (1963). The method is now usually credited to Marquardt.

Marquardt introduced a new parameter λ into the normal equations:

$$(\mathbf{J}^T\mathbf{W}\mathbf{J} + \lambda\mathbf{D})\,\Delta\mathbf{p} = \mathbf{J}^T\mathbf{W}\,\Delta\mathbf{y}$$

The parameter λ is multiplied by the diagonal matrix $\mathbf{D}$ and added to the normal equations matrix. In other words, λD_i is added to the diagonal elements of $\mathbf{J}^T\mathbf{W}\mathbf{J}$. The $\mathbf{D}$ matrix may be either the identity matrix, or it may consist of the diagonal elements of $\mathbf{J}^T\mathbf{W}\mathbf{J}$. In cases of divergence λ is then adjusted so as to ensure that the sum of squares decreases.

Now, as $\lambda\mathbf{D}$ becomes much larger than the elements of $\mathbf{J}^T\mathbf{W}\mathbf{J}$, the shift vector $\Delta\mathbf{x}$ tends to point in the direction of steepest descent and the length of the shift vector decreases in inverse proportion to the magnitude of λ:

$$\Delta\mathbf{p} \approx (1/\lambda)\mathbf{D}^{-1}\mathbf{J}^T\mathbf{W}\,\Delta\mathbf{y}$$

This is illustrated in Figure 4.7. The model used was the same as that taken for Figure 4.1, that is $y_i^{\text{calc}} = p\,e^{-qx}$. This model has the property that the Jacobian has a column of zeros when $p = 0$. The normal equations matrix is singular along the q axis. Therefore attempting to start a refinement from a point close to the q axis is asking for trouble. In the figure, the unmodified shift vector gives a point with a huge sum of squares (S increases rapidly when q becomes negative and the exponent in the model becomes positive). Choosing $\mathbf{D}$ to be a unit matrix, or simply adding the Marquardt parameter to the diagonal elements of the normal equations matrix, we can see both the rotation towards the steepest descent vector and the reduction in length. With $\lambda = 10^{-4}$ only a small improvement occurs. At 10^{-3} there is some improvement, but there is still divergence. With $\lambda = 0.01$ there is strong convergence, better in fact than with the larger λ value of 0.1. At a value of 1 the shift vector has become almost coincident with the steepest descent vector and is about

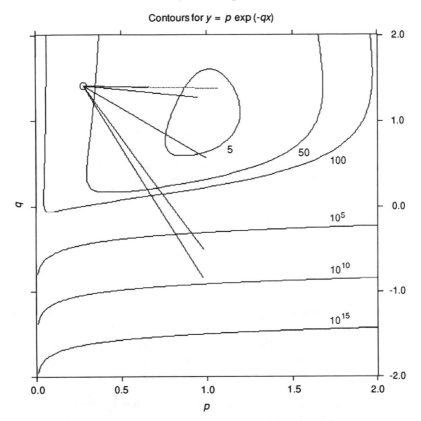

Figure 4.7. The effect of the Marquardt parameter, λ, on the direction and length of the shift vector. The model is the same as that used for Figure 4.1. The shift vector is shown for $\lambda = 10^{-4}$, 10^{-3}, 10^{-2}, 0.1 and 1. The value of λ_c for this model is 5.7×10^{-3}

half of its length. There is obviously an optimum value for λ that brings about the greatest reduction in the sum of squares.

With this illustration we can clearly appreciate how the single parameter λ, now known as the Marquardt parameter, can be varied so as to ensure that the refinement is always convergent. Moreover, as the minimum is approached the Marquardt parameter can be allowed to assume the value of zero, so that the refinement becomes a standard least-squares refinement. In this way the single parameter λ can be varied from one refinement cycle to the next to protect against divergence, without affecting the refinement unless divergence occurs and without affecting the ultimate convergence.

The general strategy for adjustment of the Marquardt parameter is as follows. Initially it is set to zero. If divergence occurs λ is set to a value that we may call the cut-off value, λ_c. A new value of S is calculated, and if this is still larger than the value on the previous cycle λ is increased. If S decreases the Marquardt parameter is passed to the next refinement cycle, and decreased if possible. Ultimately it will be decreased to less than the cut-off value, in which case it is set to zero again.

4.5.3.1 Calculation of the Cut-off Value, λ_c

Marquardt originally proposed that λ should be increased from zero in graduated steps. This may involve many calculations of S before λ begins to have a significant effect. To reduce the number of preliminary evaluations of the objective function, R, Fletcher (1971) has proposed the following choice of the cut-off value:

$$\lambda_c = 1/\text{trace}(\mathbf{J}^T\mathbf{W}\mathbf{J})^{-1}$$

(the trace of a matrix is the sum of its diagonal elements). This choice was made on the basis that it may be expected to lead to a halving of the shift vector, and Fletcher gave some arguments to show that it is usually realistic while being on the small side and therefore fail-safe. In Figure 4.7 the shift vector calculated with λ_c shows a marked improvement compared with the unmodified shift vector, although being less than optimal.

Singular value analysis (Section 3.8.2) is particularly helpful in understanding the way the Marquardt parameters works. With the decomposition $\mathbf{J}^T\mathbf{W}\mathbf{J} = \mathbf{U}\mathbf{S}\mathbf{V}^T$, the shifts are calculated as follows:

$$\Delta q_k = d_k/S_{kk} \qquad (\Delta\mathbf{q} = V\,\Delta\mathbf{p}, \qquad \mathbf{d} = \mathbf{U}^T\,\Delta\mathbf{y})$$
$$\Delta\mathbf{p} = \mathbf{V}^T\,\Delta\mathbf{q}$$

Lawson and Hanson (1974) show that when the Marquardt parameter is not zero each shift in q is multiplied by a factor that depends on λ:

$$\Delta\mathbf{q}(\lambda) = \Delta\mathbf{q}(\lambda = 0)\,\frac{S_k^2}{S_k^2 + \lambda^2}$$

Thus, when $\lambda = S_k$, Δq is halved. This provides the motivation for choosing a cut-off value for λ equal to the smallest singular value of the Jacobian; Fletcher's criterion was based on a bound for this value which can be calculated without performing the full singular value analysis. All bounds underestimate the smallest singular value.

4.5.3.2 Strategy for Increasing or Decreasing λ

A very effective strategy has been also proposed by Fletcher (1971). It is based on a prediction of the reduction that might be expected in the sum of squares.

S is expanded as a Taylor series about its current value, which is a way of linearizing the model and making S into a quadratic function of the parameters:

$$S = S_0 + \sum_j \frac{\partial S}{\partial p_j} \Delta p_j + \frac{1}{2} \sum_j \sum_k \frac{\partial^2 S}{\partial p_j \, \partial p_k} \Delta p_j \, \Delta p_k$$

$$S - S_0 = (\partial S / \partial \mathbf{p}^T) \, \Delta \mathbf{p} + \Delta \mathbf{p}^T (\partial^2 S / \partial \mathbf{p}^2) \, \Delta \mathbf{p}$$

Since $S = \mathbf{r}^T \mathbf{W} \mathbf{r}$ and $\mathbf{r} = \mathbf{y}^{obs} - \mathbf{y}^{calc}$,

$$\partial S / \partial \mathbf{p} = -2 \mathbf{J}^T \mathbf{W} \mathbf{r} \qquad \text{and} \qquad \partial^2 S / \partial \mathbf{p}^2 = 2 \mathbf{J}^T \mathbf{W} \mathbf{J}$$

Therefore the predicted reduction in the sum of squares is

$$S - S_0 = -2 \mathbf{b}^T \, \Delta \mathbf{p} + \Delta \mathbf{p}^T \mathbf{A} \, \Delta \mathbf{p}$$

where the normal equations are $\mathbf{A} \, \Delta \mathbf{p} = \mathbf{b}$. Thus, the difference between the sum of squares at arbitrary points $\mathbf{p}$ and $\mathbf{p} + \Delta \mathbf{p}$ is easy to calculate from the shift vector and the normal equations coefficients. For a non-linear model this is the approximate reduction predicted for one refinement cycle. (If the model were truly linear the prediction would be exact.)

If the actual reduction in the sum of squares is within a certain fraction of the predicted value it is probably safe to decrease the Marquardt parameter to half its value. In any case it can always be increased again if this results in divergence. Likewise, if an increase is within a certain fraction of the predicted reduction, the Marquardt parameter may be increased to double its value. For more severe cases of divergence it may save time to increase it to ten times its value. Fletcher proposed that λ be decreased if the actual reduction was more than 0.75 of the predicted, be unchanged between 0.25 and 0.75, and be increased if the actual reduction was less than 0.25 of the predicted reduction. He also proposed a simple formula whereby the factor by which λ was changed could be calculated. This has proved to be a very good compromise in practice.

There is little point in trying to find the optimum value for the Marquardt parameter in any one cycle since calculating S can be a very time-consuming process. It is better to try to keep λ as small as possible by reducing it whenever that seems to be opportune. Fletcher's strategy seems to offer a good compromise in terms of trying to keep the total length of the calculation down to a minimum.

4.6   CONVERGENCE CRITERIA AND FAILURE TO CONVERGE

The iterative refinement process consists of taking a set of initial parameters and then repeating the following process. First calculate a parameter shift by solving

$$(\mathbf{J}^T\mathbf{W}\mathbf{J} + \lambda\mathbf{D})\ \Delta\mathbf{p} = \mathbf{J}^T\mathbf{W}\ \Delta\mathbf{y}$$

in such a way that the sum of squares S decreases. Then add the shifts to the parameters:

$$\mathbf{p} \leftarrow \mathbf{p} + \Delta\mathbf{p}$$

When should the iterative process be terminated? The obvious answer—when $\Delta\mathbf{p}$ is zero—is unsatisfactory in a computer calculation because the value of zero will never occur on account of rounding errors.

A good convergence criterion is that each absolute value (indicated by $|\dots|$) of each parameter shift should be at least an order of magnitude less than the standard deviation of the parameter:

$$|\Delta p_j| \leqslant 0.1\sigma_j$$

The reasoning behind this criterion is as follows. Further refinement would produce extra precision in the parameter, but the extra digits are meaningless if they are beyond the precision dictated by the errors in the experimental data.

Unfortunately this criterion is not sufficient by itself, because in severely non-linear situations the shift vector will tend to be rather small anyway. The severely non-linear situations arise when the parameters fall in a region where the linear approximation is very poor. In these circumstances the refinement may be converging very slowly, and would hence be terminated prematurely.

Another criterion would be that the sum of squares should decrease by less than a specified amount, for example

$$|\Delta S| \leqslant 0.0001 S$$

This criterion is open to the same objection of premature termination in the case of a very slow convergence. Moreover, the figure of 0.0001 is an arbitrary one, though it has proved reasonable in many cases. There is a connection between these two criteria since, as shown in Section 4.5.3,

$$\Delta S \approx -2\Delta\mathbf{y}^T\mathbf{W}\mathbf{J}\ \Delta\mathbf{p} + \Delta\mathbf{p}^T\mathbf{J}^T\mathbf{W}\mathbf{J}\ \Delta\mathbf{p}$$

This relation may also be applied as a convergence criterion as the approximation should be good in the region of the true minimum.

It is probably safest to stop iterating only when all three criteria are satisfied. This will still not guarantee that premature termination will not occur, but it will slightly lessen the chance of it happening. The fact is that there is no universally applicable convergence criterion that can be applied to non-linear refinements.

Another consideration is that the Marquardt parameter should be zero on convergence. If this is not the case the refinement may have been terminated prematurely. This may typically occur if the Marquardt parameter becomes very large, making the parameter shifts very small and in the direction of steepest descent. In cases like this, and in the general case of very slow convergence, it is probably sensible to place a maximum limit on the number of iterations that may be performed. If this limit is exceeded the refinement is deemed to have failed.

There are a number of other ways in which a refinement may fail before convergence is reached. One is that the normal equations matrix may appear not to be positive definite. This will manifest itself in the Choleski factorization (Section 3.5.2):

$$(\mathbf{J}^T\mathbf{W}\mathbf{J}) = \mathbf{LDL}^T$$

when an element of $\mathbf{D}$, D_k say, is not positive. This failure can be remedied by introducing an arbitrary, large Marquardt parameter and restarting the refinement, which is equivalent to starting the refinement off in the direction of steepest descent.

The parameters may be given initial values, or may assume values for which the linear approximation is very bad, in which case the refinement may become very slow or even erratic. The remedy in these cases is to improve the initial parameter estimates until the refinement becomes well behaved. If good parameter estimates cannot be found, this suggests that the refinement is based on an erroneous model.*

Another way in which the refinement may fail is when a parameter refines to an unreasonable value, for example a negative value when the parameter corresponds to a physical quantity that should always be positive. A variation on this failure is the situation when a relative standard deviation is excessively large. A reasonable criterion for this is

$$\sigma_j > 2.5 \, | \, p_j \, |$$

meaning that the parameter is almost indistinguishable from zero. Both of these failures imply that the model is wrong, but do not by themselves indicate why the model is wrong.

In some cases the refinement may fail because the computer cannot handle the arithmetic required. For example, failure will occur on attempting to take the square root or logarithm of a negative number. Obviously these operations

---

* The occurrence of slow or erratic refinement can also be approached via the mathematical theory of chaos. Since the essence of the refinement is the iteration $\mathbf{p} \leftarrow \mathbf{p} + (\mathbf{J}^T\mathbf{W}\mathbf{J})^{-1}\mathbf{J}^T\mathbf{W}\,\Delta\mathbf{y}$ and the Jacobian is a function of the parameters, it follows that the evolution of the parameter set may, in some circumstances, become chaotic. By way of illustration the reader may care to try the iteration $p \leftarrow 3p\,(1 - p)$ from an initial value just less than the solution $p = \frac{2}{3}$; this refinement is exceedingly slow. The iteration $p \leftarrow 4p\,(1 - p)$ is chaotic from any initial value just less than the solution $p = \frac{3}{4}$. The Marquardt parameter damps out chaotic behaviour, but with large values it does so at the expense of rotating the shift vector to the direction of steepest descent and shortening it to the point where the refinement makes virtually no progress: chaos is replaced by stasis!

can be checked and cause an error message to appear rather than the simple computer error. The advantage of this is that the program continues to run, and the message of where the failure occurred may help to identify the cause.

Another type of failure is generated by the finite number capacity of the computer. Addition, subtraction, multiplication, division and exponentiation may lead to 'overflow'. The arithmetic operations are too numerous to check each one individually so these failures are difficult to deal with. One instance that has occurred frequently is that the Marquardt parameter has been doubled repeatedly so many times that it overflows. This can be checked, and if this failure occurs it indicates that new parameter estimates are required.

The main point to realize is that all failures to converge originate with either bad parameter estimates or an incorrect model. Indeed, failure to converge from a variety of parameter estimates is a sure sign that the model is deficient in some respect. It may have too many or too few parameters.

4.7

4.7.1 NEWTON'S METHOD

Newton's method provides an alternative method of solving a non-linear least-squares problem. It is based on a different method of linearization. The sum of squares S is expanded as a Taylor series in the parameters, up to second order. This is equivalent to assuming that S is quadratic in the parameters, which in turn is equivalent to assuming that y^{calc} is linear in the parameters:

$$S = S^0 + \sum_{j=1,n} \frac{\partial S}{\partial p_j} p_j + \frac{1}{2} \sum_{j=1,n} \sum_{k=1,n} \frac{\partial^2 S}{\partial p_j \, \partial p_k} p_j p_k$$

$$S = S_0 + \mathbf{p}^{\mathrm{T}} \mathbf{b} + \tfrac{1}{2} \mathbf{p}^{\mathrm{T}} \mathbf{H} \mathbf{p}$$

$\mathbf{H}$ is the so-called Hessian matrix of second derivatives. To minimize S we differentiate it with respect to the parameters:

$$\mathbf{g} = \mathbf{b} + \mathbf{H} \mathbf{p} \qquad (g_j = \partial S / \partial p_j, j = 1, n)$$

At the minimum in S the gradient $\mathbf{g}$ must be zero and the parameters must have values $\mathbf{p}^{min}$, so

$$\mathbf{b} = -\mathbf{H} \mathbf{p}^{min}$$

Therefore the general equation for the gradient becomes

$$\mathbf{g} = -\mathbf{H} \mathbf{p}^{min} + \mathbf{H} \mathbf{p} = \mathbf{H} (\mathbf{p} - \mathbf{p}^{min}) = -\mathbf{H} \, \Delta \mathbf{p}$$

A second condition for the minimum is that the Hessian matrix must be positive definite.

When S is the sum of squares $\mathbf{r}^T\mathbf{W}\mathbf{r}$, where $\mathbf{r} = \mathbf{y}^{obs} - \mathbf{y}$, we can differentiate it with respect to the parameters:

$$\frac{\partial S}{\partial p_j} = \frac{\partial S}{\partial r_j}\frac{\partial r_j}{\partial p_j} = -2\mathbf{J}^T\mathbf{W}\mathbf{r}; \qquad J_{ij} = \frac{\partial y_i}{\partial p_j}$$

$$\frac{\partial^2 S}{\partial p_j \partial p_k} = 2\mathbf{J}^T\mathbf{W}\mathbf{J} - 2\sum_{i=1,m}\sum_{h=1,m}\frac{\partial^2 y_i}{\partial p_j \partial p_k}W_{ih}r_h$$

If we were to ignore the second term in the last equation and insert the expressions for the derivatives into the equation $\mathbf{g} = -\mathbf{H}\,\Delta\mathbf{p}$ we would obtain the normal equations of linear least squares. This happens because in that case the second derivatives of y with respect to the parameters are zero by the definition of a linear model.

If we define a non-linear model such that the second derivatives of y with respect to the parameters are not zero we could use the full expression for the Hessian given above. Because the model is now non-linear, solution of $\mathbf{g} = -\mathbf{H}\,\Delta\mathbf{p}$ will not generally locate the minimum. The calculation thus becomes iterative. Newton first expounded the basic method and Raphson formulated it as an iterative method. Hence we may call this the Newton–Raphson method.

From the point of view of least-squares calculations the Newton–Raphson method has little in its favour. The main advantage that could be ascribed to it is that ultimate convergence is faster than with the standard (Gauss) method. However, this will save only a few iterations, as the Gauss method converges quite rapidly near the minimum. Set against the improved speed of convergence is the time required to compute the second derivatives, of which there are mn^2 in number. A more serious disadvantage is that the Hessian is not guaranteed to be positive definite, though there are instances in which it is so. The Hessian in the non-linear problem is not constant, so we can see that it will be positive definite in the region of a minimum and, if the initial point is close to the minimum, the method should converge to that minimum.

In a few cases tests have been carried out to compare the Newton–Raphson method with the Gauss method, and it has been found to give little benefit. The main area of application is in the solution of simultaneous non-linear equations. This can be seen as a special case of least-squares minimization in which S is zero at the minimum. However, even in this case use of the Gauss method may be satisfactory. When solving n equations for n parameters the Jacobian matrix is square. Therefore the solution to the normal equations

$$\Delta\mathbf{p} = (\mathbf{J}^T\mathbf{W}\mathbf{J})^{-1}\mathbf{J}^T\mathbf{W}\,\Delta\mathbf{y}$$

becomes

$$\Delta\mathbf{p} = \mathbf{J}^{-1}\mathbf{W}^{-1}\mathbf{J}^{T-1}\mathbf{J}^T\mathbf{W}\,\Delta\mathbf{y} = \mathbf{J}^{-1}\,\Delta\mathbf{y} \qquad \text{or} \qquad \mathbf{J}\,\Delta\mathbf{p} = \Delta\mathbf{y}$$

This is somewhat simpler than the standard set of normal equations, and only becomes more complicated if **J** is not symmetrical.

4.7.2   THE DAVIDON–FLETCHER–POWELL (DFP) METHOD

This is a method belonging to the class of pseudo-Newton methods, which are characterized by using an updating formula for the *inverse* of the Hessian. With an approximate inverse Hessian the parameter shifts are calculated as

$$\Delta \mathbf{p} = -\mathbf{H}^{-1}\mathbf{g}$$

The updating formula used in the DFP algorithm is a rank 2 formula using information from first derivatives only. It has been shown that for *n* parameters the updating should result in a good approximation to the inverse Hessian in *n* iterations. Sadler (1975) gives a good summary of the DFP method.

As with the Newton method, it seems that the DFP method has little advantage over the Gauss method. True, the DFP method is faster per iteration than Newton's method, because no second derivatives are calculated, but more iterations are needed.

The author has used the DFP method for the calculation of vibrational force constants (Gans, 1975). This calculation was really the solution of a set of non-linear equations, with as many unknown force constants as observed vibration frequencies. However, the normal equations matrix of the standard least-squares method was often rather ill-conditioned with respect to inversion, and in this instance the DFP algorithm proved better in terms of always converging to a solution.

Another test of the method was made with equilibrium constant computation. This used an adaptation, due to Stewart, in which the derivatives are estimated numerically (Gans and Vacca, 1974). The standard least-squares method was found to be preferable.

4.8

4.8.1   NON-DERIVATIVE METHODS

All the methods discussed so far in this chapter use analytical expressions for the derivatives $\partial y / \partial p$. A variety of methods is available in which expressions for the derivatives are not needed. Such methods are superficially attractive insofar as they only require a formula for calculating the objective function and are therefore simpler to program. However, there is no doubt in my mind that the non-derivative methods are inferior in two respects. They consume more computer time and they have poor convergence properties.

The nub of the problem is this. Is there a model so complicated that expressions for the derivatives cannot be deduced? My belief is that such cases are few and far between. An outline of the rules of partial differentiation is given in Appendix 4. Since the model must be capable of being stated in the form $f(y, p, x) = 0$ and since there must be an expression relating y to the parameters in order that the residuals can be calculated, it follows that the derivatives can nearly always be calculated, although in some cases the calculation may be very lengthy. Perhaps the most lengthy instance occurs in finding the derivatives of integrals, which are themselves integrals. In view of the above, non-derivative methods will be discussed rather briefly.

4.8.2 GRADIENT METHODS

In the Gauss–Newton, Newton and DFP methods the derivatives may be calculated using a difference formula. With $\mathbf{y} = f(\mathbf{p}, \mathbf{x})$,

$$\frac{\partial \mathbf{y}}{\partial \mathbf{p}} \approx \frac{f(\mathbf{p} + \delta\mathbf{p}, \mathbf{x}) - f(\mathbf{p}, \mathbf{x})}{\delta\mathbf{p}}$$

The chief problem in applying this formula is to choose the size of the increments δp_j. The estimate of the derivative will be inaccurate if the increment is too large for obvious reasons, and also if it is too small, for reasons of round-off error in the computer. A better formula to use is the so-called central difference formula:

$$\frac{\partial \mathbf{y}}{\partial \mathbf{p}} \approx \frac{f(\mathbf{p} + \delta\mathbf{p}, \mathbf{x}) - f(\mathbf{p} - \delta\mathbf{p}, \mathbf{x})}{2\,\delta\mathbf{p}}$$

but this requires the extra evaluation of the function $f(\mathbf{p} - \delta\mathbf{p}, \mathbf{x})$. There are ways in which the error in the derivative can be estimated, so that a rational choice can be made of approximating formula and of increment size (Stewart, 1967; Lill 1970).

A simple difference formula was used in the programs LETAGROP and SCOGS (ses Chapter 10) and has proved to be satisfactory in many instances. In both those cases, however, the problem could be restated in a form such that the derivatives could be calculated analytically.

4.8.3 DIRECT SEARCH METHODS

In the simplest of these methods each parameter is varied in turn in such a way as to reduce the value of the objective function. This method is really only of historical interest, but more sophisticated methods based on a similar principle have been developed. Powell's method (1964) is based on a scheme of conjugate gradients, and is highly recommended for those systems that can be treated by the standard least-squares technique, but for which the calculation of derivatives is tedious or impossible.

Non-linear Least Squares

For some systems the standard least-squares method is not satisfactory. Perhaps the commonest such system involves models with multiple exponentials (Section 8.2.2). For such systems the simplex method of Nelder and Mead is robust and often effective; a FORTRAN coding has been published (O'Neill, 1971).

A simplex is a geometrical entity that has $n + 1$ vertices corresponding to variations in n parameters. For two parameters the simplex is a triangle, for three parameters the simplex is a tetrahedron and so forth. The value of the objective function is calculated at each of the vertices. An iteration consists of the following process. Locate the vertex with the highest value of the objective function and replace this vertex by one lying on the line between it and the centroid of the other vertices. Four possible replacements can be considered, which I call contraction, short reflection, reflection and expansion. The progress of a simplex minimization is illustrated in Figure 4.8.

The model used for this calculation was

$$y_i = e^{-px_i} + e^{-qx_i}$$

and a contour map for this model is shown in Figure 4.6. Since the model has two parameters, p and q, the simplex is a triangle. The first simplex was formed by taking $p = 0$ and $q = 0$ with an arbitrary initial step of 0.25. The evolution of the minimization can be followed by seeing which of the four options was used in obtaining each successive simplex:

2 Expansion
3 Reflection
4 Expansion
5 Reflection
6 Short reflection
7 Short reflection

At this stage the vertex $p = 0.75$, $q = 0.75$ has been found, which happens to lie in the valley in the contour map. There then follows a series of contractions (simplexes 8 and 9) and reflections until eventually the simplexes move down the valley towards the minimum at $p = 0.5$, $q = 1.5$.

This illustration exemplifies the salient characteristics of the simplex method. It starts with an arbitrary simplex. Neither the shape nor position of this are critically important, except insofar as it may determine which one of a set of multiple minima will be reached. The simplex then expands or contracts as required in order to locate a valley if one exists. Then the size and shape of the simplex is adjusted so that progress may be made towards the minimum. Note particularly that if a pair of parameters are highly correlated, *both* will be simultaneously adjusted in about the correct proportion, as the shape of the simplex is adapted to the local contours. For further details the reader is referred to O'Neill (1971).

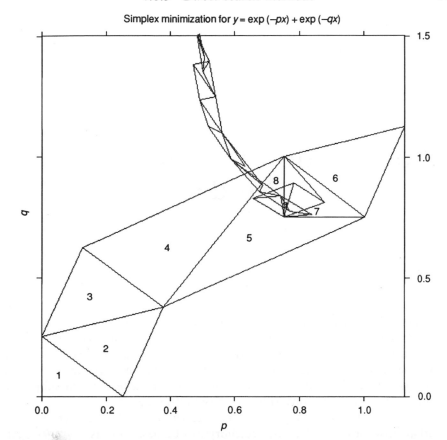

Figure 4.8. Progress of the Nelder–Mead simplex minimization operating on the model $y = e^{-px} + e^{-qx}$. Contours of the objective function for this model are shown in Figure 4.6

For a system such as the double exponential the simplex method can be used to find good estimates for the least-squares parameters. Unfortunately it does not provide estimates of the parameter errors, etc. It is therefore to be recommended as a method for obtaining initial parameter estimates that can be used in the standard least-squares method.

5

Formulation and Selection of Models

5.1 EMPIRICAL AND THEORETICAL MODELS

The method of least-squares requires that the experimentalist defines a model with which to fit the data. This is both a strength and a weakness. The strength of the method lies in its ability to furnish values of the parameters, which may have scientific significance, of the chosen model. The weakness lies in the fact that it is a model-building technique, and as such is dependent on subjective judgements as to what constitutes the 'correct' model.

Models can be classified into two main categories, although there is some overlap between them. The first category can be described as *empirical*. An empirical model is one that has no foundation in theory and is simply being used to reduce the data to a very compact form. The most common example is when we think that the data show a linear dependence on some variable, and we fit a straight line. A typical instance would be in the construction of a calibration curve. Then the objective is to reduce the calibration data to an expression containing one or two parameters in order to back-calculate the value of the variable from a measurement of a physical property. Going one step further with this example, it may be that the actual measurements depart from linearity for some reason, but that the departure is reproducible. In that case we would model the data with a curve based on a simple mathematical function. As long as the objective of the data-fitting is a purely empirical one, it does not matter what function is chosen. The only requirement is that the model gives a good representation of the experimental data.

A good example is provided by colorimetry or atomic absorption spectrophotometry. Theory suggests that in ideal circumstances a plot of absorbance against concentration should be a straight line passing through the origin. However, experience shows that even when the relationship between absorbance and concentration is linear, the best straight line rarely passes through the origin. It is perfectly acceptable to fit a general straight line for the purpose of determining concentration from absorbance. Moreover, because of departures from ideality the calibration plot may be curved in a way that theory cannot handle. As long as the calibration plot is reproducible it does not matter what shape of curve is used to represent it.

Typical functions that have been used to fit data empirically include polynomials, rational functions (quotients of polynomials), exponential, logarithmic and trigonometric functions. When using one of the latter care should be taken over the choice of weights, since any transformation of the observations may have a significant effect on errors. Transformations to linear models are discussed in Section 7.6.4.

When the model has a theoretical basis the objectives are different from when the model is empirical. It may be that the model is known with great certainty. If that is so the objective is to determine the parameters of the model in order to gain theoretical knowledge. A more difficult situation is the one in which the model is not known with certainty; indeed, there may be two or more candidate models and the objective is to choose between them.

Part of the process of building the model is to designate which parameters are to be determined. Some parameters may be 'known' from other experiments; it will have to be decided if these parameters are to be refined, i.e. redetermined, or held fixed to the known values. For example, in the determination of the dissociation constant of a weak base it is necessary to know the self-dissociation constant of water K_w for the conditions of temperature and ionic strength being used in the determination. K_w is best determined in a separate experiment, so that it becomes a fixed parameter in the more complicated calculation. If any parameters are linked by a theoretical relationship constraints should be used as described in Section 5.2. Indeed, if any two parameters p and q are related by a linear expression, $p = aq + c$, it is essential to use constraints since otherwise the normal equations matrix will be singular, and refinement is impossible.

Since the object of building a model is to obtain a good fit to experimental data, the control of data quality is important for model selection. If the model is not adequate this may be the fault of the data and it could be that the same model would be adequate with better data. Therefore, in some instances model building and experimentation depend upon each other and a preliminary fit may provide a specification for the data quality. Data quality is discussed in Section 5.3.

Whether the model is empirical or theoretical one will need to establish whether it gives an adequate representation of the data. How does one know when the model is adequate? The answer is that we can never be certain; this is the inherent weakness of any method that requires a model to be constructed. However, there are ways in which we can assure ourselves that the model is adequate within the limits of experimental error. These are discussed in Chapter 6.

5.2

5.2.1 EQUALITY CONSTRAINTS

It often happens that some parameters in a model are related to some others. We may say that the former are *constrained* by the relationships. A typical example occurs in fitting n.m.r. or Mössbauer spectra when the bands occur as multiplets with component heights related by expressions such as $h_1 = \frac{1}{2} h_2, h_3 = \frac{1}{2} h_2$; in this instance we can write the constraint equations as

$$\left. \begin{array}{l} h_1 = \frac{1}{2} h_2 \\ h_2 = 1\, h_2 \\ h_3 = \frac{1}{2} h_2 \end{array} \right\}, \qquad \begin{bmatrix} h_1 \\ h_2 \\ h_3 \end{bmatrix} = \begin{bmatrix} \frac{1}{2} \\ 1 \\ \frac{1}{2} \end{bmatrix} \begin{bmatrix} h_2 \end{bmatrix}$$

Clearly, if we know h_2 we can calculate h_1 and h_3. However, the model contains all three parameters. We can eliminate the constrained parameters by replacing them by expressions that contain only the unconstrained parameters. Using the same example,

$$y_k = \frac{h_1}{L_{1k}} + \frac{h_2}{L_{2k}} + \frac{h_3}{L_{3k}} \rightarrow \frac{h_2}{2L_{1k}} + \frac{h_2}{L_{2k}} + \frac{h_2}{2L_{3k}}$$

(L stands for the denominator of a Lorentzian). The least-squares refinement requires the partial derivative $\partial y / \partial h_2$, which is the sum of three terms, as can be seen by inspection.

A general way of constructing the Jacobian is to write

$$q_k = \sum_{j=1,n} C_{kj} p_j + a_j; \qquad \mathbf{q} = \mathbf{C}\mathbf{p} + \mathbf{a}$$

by which we mean that the constrained parameters q are linear combinations of the unconstrained parameters $\mathbf{p}$ and some constants $\mathbf{a}$. It is convenient to include in the list of constrained parameters those unconstrained parameters that are simply equal to themselves. To obtain the Jacobian for the independent parameters, we use the chain rule:

$$\frac{\partial y_i}{\partial p_j} = \sum_{k=1,n} \frac{\partial y_i}{\partial q_k} \frac{\mathrm{d} q_k}{\mathrm{d} p_j} = \sum_{k} J_{ik} \frac{\mathrm{d} q_k}{\mathrm{d} p_j}$$

$\mathbf{J}$ represents the Jacobian with respect to all parameters. The second term is obtained by differentiating the constraint equations:

$$\frac{\mathrm{d}q_k}{\mathrm{d}p_j} = C_{kj}$$

Therefore

$$\frac{\partial y_i}{\partial p_j} = J_{ij} \leftarrow \sum_k J_{ik}C_{kj}; \qquad \mathbf{J} \leftarrow \mathbf{JC}$$

The constrained problem closely resembles the unconstrained one. Firstly all the derivatives are calculated and then the derivatives for the independent parameters are obtained by combining these derivatives together linearly, with coefficients given by the equations defining the constraints. This Jacobian is used to calculate the shifts on the independent parameters—hence their new values and hence the values of the constrained parameters.

Let us now complete the illustration involving the triplet of bands. Realistically there would also be other constraints, on the widths, w, and positions, p, of the bands. For example,

$$w_1 = w_2, \qquad w_3 = w_2, \qquad p_1 = p_2 - \delta, \qquad p_3 = p_2 + \delta$$

The complete constraints matrix is then

$$\begin{bmatrix} h_1 \\ p_1 \\ w_1 \\ h_2 \\ p_2 \\ w_2 \\ h_3 \\ p_3 \\ w_3 \end{bmatrix} = \begin{bmatrix} \frac{1}{2} & 0 & 0 & 0 \\ 0 & 1 & 0 & -1 \\ 0 & 0 & 1 & 0 \\ 1 & 0 & 0 & 0 \\ 0 & 1 & 0 & 0 \\ 0 & 0 & 1 & 0 \\ \frac{1}{2} & 0 & 0 & 0 \\ 0 & 1 & 0 & 1 \\ 0 & 0 & 1 & 0 \end{bmatrix} \begin{bmatrix} h_2 \\ p_2 \\ w_2 \\ \delta \end{bmatrix}$$

The big advantage of eliminating the constrained parameters before starting the calculation is twofold: the Jacobian automatically has the correct number of columns for those parameters being refined, i.e. it has rank $n - s$, and secondly the size of the calculation has been considerably reduced—from nine to four parameters in the above example. Therefore in general linear equality constraints should always be eliminated, rather than try to perform an unconstrained minimization.

Non-linear equality constraints are more complicated to handle. The s constrained parameters must be calculated by solving the s constraint equations. An example occurs as a subproblem in the computation of equilibrium constants. The concentrations of the species in solution may be considered to be subject to a conservation law, the conservation of mass. This means that the total mass of each independent species must be constant. Consider the simplest instance, with two independent species, ligand L and hydrogen ions H. Dividing the masses by a volume we can write the constraints in terms of concentrations, denoted by [L], etc.:

$$[\text{Total}_L] = [L] + \Sigma\, a\beta_{ab}[L]^{\,a}[H]^{\,b}$$
$$[\text{Total}_H] = [H] + \Sigma\, b\beta_{ab}[L]^{\,a}[H]^{\,b}$$

The two constrained parameters [L] and [H] are found by solving these two equations using the current values of the equilibrium constants. By always adjusting the constrained parameters so that the conservation law is obeyed these parameters are eliminated as variables in the least-squares refinement. The derivatives of the unconstrained parameters β_{ab} are later obtained by using these values, as discussed in Chapter 10.

The general procedure followed in this example consists of two phases: finding the values of the constrained parameters and using these values to form the Jacobian. Although the parameters are eliminated from the least-squares refinement as far as finding their values is concerned, they are not to be regarded as constant when forming the Jacobian with respect to the unconstrained parameters.

5.2.2   INEQUALITY CONSTRAINTS

Before looking at ways of dealing with inequality constraints we must ask a fundamental question: are they necessary? In the physical sciences and in least-squares minimizations in particular, inequality constraints are not always justified. The most common inequality constraint is that some number that relates to a physical quantity should be positive, $p_j > 0$. If an unconstrained minimization leads to a negative value, what are we to conclude? There are three possibilities: (a) the refinement has converged to a false minimum; (b) the model is wrong; (c) the parameter is not well defined by the data and is not significantly different from zero. In each of these three cases a remedy is at hand that does not involve constrained minimization: (a) start the refinement from good first estimates of the parameters; (b) change the model; (c) improve the quality of the data by further experimental work. If none of these remedies cure the problem of non-negativity constraints, then something is seriously wrong with the patient, and constrained minimization will probably not help.

All inequality constraints can be reduced to the form of non-negativity constraints. For example,

$$f(\mathbf{p}) \geqslant \text{constant} \equiv f(\mathbf{p}) - \text{constant} \geqslant 0$$
$$f(\mathbf{p}) \geqslant g(\mathbf{p}) \qquad \equiv f(\mathbf{p}) - g(\mathbf{p}) > 0$$

The simplest way to deal with these constraints (if it is absolutely unavoidable) is by a change of variable. The following examples illustrate this point:

$$q = p^2 \text{ or } q = e^p \text{ ensure that } q \geqslant 0$$
$$q = \sin p \text{ ensures that } -1 \geqslant q \leqslant 1$$
$$q = p + z^2 \text{ ensures that } q \geqslant p$$

However, these changes of variable may have a serious disadvantage. If the model is linear in a parameter, it is non-linear in its square root (or logarithm or arcsine). This will cause the refinement to need more iterations and may also create multiple minima. Therefore the change of variable should be used with caution.

The example quoted in Section 5.2.1

$$[\text{Total}_L] = [L] + \Sigma \, a\beta_{ab}[L]^a[H]^b$$
$$[\text{Total}_H] = [H] + \Sigma \, b\beta_{ab}[L]^a[H]^b$$

is one in which it was deemed essential that the variables [L] and [H] should be positive. (Apart from anything else it is necessary at a later stage to take the logarithm of [H].) These equations have multiple solutions, as many solutions as the largest sum of a and b. Moreover, there are situations in which it is virtually impossible to get good initial estimates for both [L] and [H]. Changing the variables to $[L] = p^2$ and $[H] = q^2$ does permit one to find the unique solution in which both [L] and [H] are positive.[*] However, the cost is that the refinement may take twice as much time.

Constraints may be introduced, on an *ad hoc* basis, if the unconstrained minimization would cause them to be violated. For example, if the parameter shifts are such that the new parameters would violate a non-negativity constraint, the shift vector can be reduced in length so as to reduce the largest element to -0.9 times the corresponding parameter value; that parameter would then be reduced to a tenth of its former value. Of course the difficulty with *ad hoc* methods is that they do not come with guarantees, and they may fail at any time.

Mathematicians have devoted some considerable time and effort to the problem of constrained minimization, as the interested reader will find from the references. Outside the physical sciences the problem is of major importance. However, we will end this section as we began, by emphasizing that if the constraints can be avoided by reformulating the model then that should be done.

5.2.3 THE METHOD OF PREDICATE OBSERVATIONS

This procedure is one in which some parameters are loosely constrained and will be of use wherever there is prior information concerning the probable

---

[*] Such a solution exists if all the equilibrium constants are positive.

values of the parameters. It is particularly applicable to situations in which a molecular structure can be predicted with some degree of confidence by extrapolation from known structures of similar molecules. Thus, bond lengths and inter-bond angles, and quantities derived from them, such as moments of inertia, can often be estimated with high precision. Vibrational force constants is another case in point (Bartell *et al.*, 1975).

The basis of the method is that the predicted value of a parameter is considered as an additional *observation*. However, to make the method work the predicate observation must be assigned a realistic weight corresponding to an estimate of the uncertainty in the prediction. Thus, the method is used in the context of a weighted least-squares calculation. The Jacobian, weight matrix and observations can be written down in a partitioned form:

$$\mathbf{J} = \begin{pmatrix} \mathbf{J} \\ \mathbf{J}_p \end{pmatrix}, \qquad \mathbf{W} = \begin{pmatrix} \mathbf{W} & | & \mathbf{0} \\ \mathbf{0} & | & \mathbf{W}_p \end{pmatrix}, \qquad y = \begin{pmatrix} \mathbf{y}^{\text{obs}} \\ y_p \end{pmatrix}$$

The items with a p subscript refer to the predicate observations. $\mathbf{W}_p$ is diagonal as there cannot be correlation between predicate observations. The normal equations will take the form

$$(\mathbf{J}^{\mathsf{T}}\mathbf{W}\mathbf{J} + \mathbf{J}_p^{\mathsf{T}}\mathbf{W}_p\mathbf{J}_p)\ \Delta\mathbf{p} = \mathbf{J}^{\mathsf{T}}\mathbf{W}\ \Delta\mathbf{y} + \mathbf{J}_p^{\mathsf{T}}\mathbf{W}_p\ \Delta\mathbf{y}_p$$

$\mathbf{J}_p$ has as many rows as there are predicate observations, and this is not necessarily the same as the total number of parameters. Therefore the effect of introducing the predicate observations is as follows. To those rows and columns of the normal equations matrix $\mathbf{J}^{\mathsf{T}}\mathbf{W}\mathbf{J}$ that refer to a predicated parameter are added the corresponding rows and columns of $\mathbf{J}_p^{\mathsf{T}}\mathbf{W}_p\mathbf{J}_p$. Inclusion of the predicate observations is equivalent to adding to the model one equation for each predicated parameter of the form

$$y = p$$

so that the elements of $\mathbf{J}_p$ are simply 0 or 1. Therefore the net effect on the normal equations is simply to add $W_{p,j}$ to the diagonal elements $(\mathbf{J}^{\mathsf{T}}\mathbf{W}\mathbf{J})_{jj}$ for each of the predicated parameters p_j. On the right-hand side, $W_{p,j}\ \Delta y_{p,j}$ is added to appropriate rows.

The main advantage claimed for the method of predicate observations is that it improves the condition of the normal equations matrix by reducing the correlation coefficients between parameters. This comes about through the additions to the diagonal elements of the normal equations matrix. However, any results obtained by this method are only as good as the values of the predicted parameters and their uncertainties.

Another advantage of the method is that it constrains certain parameters to a restricted range of values. This could be useful in general least-squares problems as a means of stopping the parameter values from changing too drastically, especially at the start of a refinement. Nevertheless, it should always be borne in mind that one is no longer fitting purely experimental data.

5.3 CONTROL OF DATA QUALITY

It is a truism to say that the quality of information one obtains from a set of experiments is only as good as the quality of data allows. All the sophisticated computer techniques that are available only permit one to extract from the data the maximum possible amount of information that is contained in it. Therefore, if one is seeking high-quality information one must start with high-quality data.

The control of data quality begins with the choice of instrumentation. The instrument's specifications determine the ultimate data precision that can be achieved. The instrumental settings determine the actual precision in any one experiment. The accuracy of results is determined by the experimenter's ability to control systematic errors.

What does the term 'high quality data' mean? It means data with low systematic and random errors.

The systematic errors can only be reduced by careful experimentation and the use of quality-assured standards, such as are provided by national institutions (e.g. the National Bureau of Standards or National Physical Laboratory). To reduce systematic error one needs to be aware of all the factors that might affect the experimental results. Sometimes a factor may be unsuspected until after the experiment has been performed and the data analysed. In that case it is better to repeat the experiments with that factor under control than to try to extract information from flawed data.

Sometimes the instruments used to make the measurements are themselves designed to reduce systematic error. For example, a double-beam spectrophotometer automatically removes the effects of variation in source intensity during an experiment, whereas a single-beam instrument cannot do so. Some manufacturers of single-beam instruments try to compensate for this by stabilizing the source intensity, but others do not—it usually boils down to a question of cost as to whether it is done. Some experiments, such as FTir, fluorescence or Raman spectroscopy, are intrinsically single-beam experiments, and in these cases one should ideally monitor the source intensity, and correct for variations—a procedure that is unfortunately rarely carried out.

Systematic errors are also reduced whenever measurements are made against a calibrant, rather than by absolute value. This is commonplace in chemical analysis. Ideally the calibration conditions should be as close as possible to the experimental. For example, to measure the quantity of mercury in seawater the calibration mixtures should be made up from known amounts of mercury in a mixture whose composition closely mimics the local composition of seawater.

As we have seen in Section 2.4, when one thinks about reducing experimental error, one knows that reduction of the largest error pays the largest dividend. This applies as much to systematic errors that affect the accuracy of a result as to random errors that affect its precision. However, the magnitude of systematic errors is extremely difficult to estimate. Perhaps the only way it can be achieved is by having the experiments repeated in different laboratories, in different countries, with different apparatus and by different people. A study of the determination of the formation constants of glycine complexes of nickel has clearly demonstrated that inter-laboratory accuracy differences are larger than any individual precision (Braibanti *et al.*, 1987). Errors that are systematic to one laboratory become random when compared to the errors from other laboratories.

Random errors should always be estimated, though the estimation is sometimes far from easy. Most instruments come with precision specifications that give one an idea of the instrumental errors, but all too frequently the specifications are inadequate. If all else fails we can estimate the precision of an instrument as a fraction of the least significant digit in a typical measurement.

Repeated experiments offer a means of estimating precision experimentally. This estimate may apply directly to the results, thus bypassing estimation of error in individual measurements. That is why chemical analyses should be carried out in duplicate or triplicate. The true result of a series of replicate experiments is not a single number but rather a range within which the result is known to lie. This range may be expressed in terms of probability, confidence limits, standard deviations or any other criterion.

One form of repeated measurement that merits further discussion is the method of *coaddition*. In this method an attempt is made to repeat a measurement exactly. Perhaps the best-known example is Fourier transform nuclear magnetic resonance spectroscopy (FTn.m.r.). In this technique the sample in a magnetic field is subject to a pulse of electromagnetic radiation and the radiation intensity is measured as a function of time, providing the so-called free induction decay (f.i.d.). A typical f.i.d. consists of exponential and sinusoidal functions multiplied together, which has been typically sampled at, say, 1000 equally spaced time intervals (measured in milliseconds). In the coaddition process the data are stored in a computer, and the sample is subjected to another pulse, producing a second f.i.d. Assuming that each measurement of f.i.d. intensity is taken *at the same elapsed time* the data can be added together and/or averaged. By continuing the sequence a large number of data sets may be combined together. The assumption is important. It is based on the idea that the true value of the measurement of the f.i.d. at a given time is a constant, and that the measured value is equal to that constant value plus a random experimental error.

If the error on a measurement at time t is a random variable with mean value zero and standard deviation σ_t we can estimate the value y_t of the measurement. Suppose that we have taken m measurements. We can assume that each measurement has equal weight, as this is equivalent to assuming that the measurement has constant variance. It follows that the least-squares estimate of y_t is the mean of all the values:

$$y_t^{\text{estimate}} = \sum_{k=1,m} (1/m)y_{kt}^{\text{obs}} = \bar{y}_t$$

It follows from the law of propagation of error that the expectation value of the variance σ_m^2 on the mean is given by

$$\sigma_m^2 = \sum_{k=1,m} (1/m)^2 \sigma_t^2 = \sigma_t^2/m$$

$$\sigma_m = \sigma_t/\sqrt{m}$$

Thus, the expected value of the standard deviation on the mean of m measurements is inversely proportional to the square root of m, the number of measurements. This is an important result in its own right. It allows us to estimate the amount by which the error is reduced as a result of coaddition. For example, in FTn.m.r. if one scan takes 10 ms, 10 000 scans may be accomplished in a minute and 40 seconds, with an improvement of 100 in the signal-to-noise ratio. This is the basis on which spectra may be obtained on ^{13}C at natural abundance (1.108%).

One may use the experimental data to obtain the so-called sample variance, s_t^2:

$$s_t^2 = \frac{\Sigma (y_{kt} - \bar{y}_t)^2}{m-1}$$

$$s_t = \sqrt{\left\{\frac{[\Sigma y_{kt}^2 - (\Sigma y_{kt})^2]/m}{m-1}\right\}}$$

Although the sample variance is not generally calculated in FTn.m.r., it will be useful in other cases where data are coadded, as it provides a way in which the weights may be calculated when fitting the data by a least-squares procedure.

An example is shown in Figure 5.1. This shows part of the infrared spectrum of polystyrene recorded on a scanning spectrophotometer. Thirty-six individual scans were recorded; the mean of these should have errors which are one-sixth of the errors in a single scan. Each scan was drawn on the plot. The error estimates at each point have also been plotted. Although the error estimate is a bit noisy it clearly shows that the magnitude of the error depends on the magnitude of the signal; closer inspection revealed that the dependence was exponential.

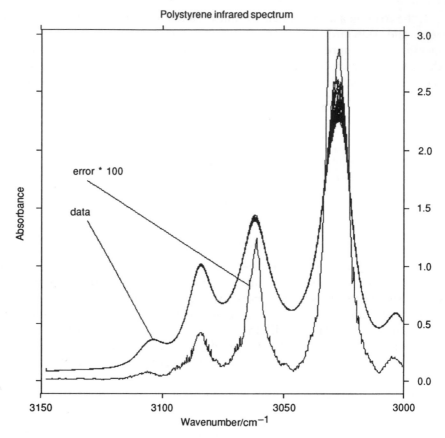

Figure 5.1. Data from 36 scans of part of the infrared spectrum of polystyrene (upper curves) and the standard deviations at each wavenumber (lower curve)

Let us now consider what the error might be on the estimate of the sample standard deviation, s_t. To do this we must make some assumption concerning the nature of the probability distribution of the errors on the measurements made at time t. Let us assume that they belong to a normal distribution of mean zero and variance σ_t^2. It can be shown that the quantity ms_t^2/σ_t^2 belongs to a χ^2 probability distribution with $m - 1$ degrees of freedom. From this we may calculate the probability that the ratio s_t/σ_t will lie in a given range of values. Mandel (1964) describes the procedure in detail. Some typical results are shown in graphical form in Figure 5.2.

In this figure the probable percentage error on s_t is shown as a function of the number of data points, m, for three different probabilities. The 99% confidence level means that there is a 99% probability that the relative error will be less than or equal to the value plotted, i.e. there is a 1% probability

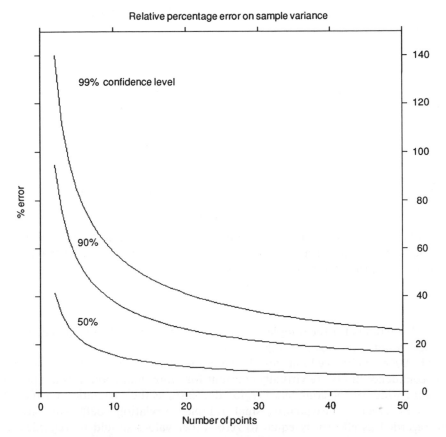

Figure 5.2. Relative percentage error on sample variance as a function of the number of measurements in the sample (see text for details of the calculation)

that it will be larger. The first thing to notice is that the relative error on s_t decreases sharply as the number of measurements increases from 2 to about 10, and then it decreases very slowly. Thus, for two measurements there is a high probability that the sample standard deviation will be out *by a factor of 2!* Even when 50 measurements of the same quantity have been taken there is a high probability that the sample standard deviation will be out by as much as 10–20% compared to the true standard deviation of the experimental errors. One may also estimate the sample covariance terms:

$$(COV)_{ij} = \sum_{k=1,m} (y_{ki} - \bar{y}_i)(y_{kj} - \bar{y}_j)/(m-1)$$

$$= \frac{\sum y_{ki}y_{kj} - (\sum y_{ki} \sum y_{kj})/m}{m-1}$$

though for statistical purposes the sample correlation coefficient, r_{ij}, is easier to handle:

$$r_{ij} = \frac{\Sigma(y_{ki} - \bar{y}_i)(y_{kj} - \bar{y}_j)}{[\Sigma(y_{ki} - \bar{y}_i)^2 \Sigma(y_{kj} - \bar{y}_j)^2]^{1/2}}$$

The probability distribution function for the sample correlation coefficient is rather complicated, but it may be approximated by a simple formula when m is 'large'. With the transformations

$$z = \frac{1}{2}\log_e\frac{1 + r_{ij}}{1 - r_{ij}}; \qquad \zeta = \frac{1}{2}\log_e\frac{1 + \rho_{ij}}{1 - \rho_{ij}}$$

where ρ_{ij} is the population correlation coefficient, the transformed variable

$$z_{ij}^* = \sqrt{(m - 2)}(z_{ij} - \zeta_{ij})$$

belongs to a standard normal distribution. The approximation is reasonable for about five points or more. Thus confidence limits for z^* may be evaluated, and by using the inverse transformation

$$r = \frac{e^{2z} - 1}{e^{2z} + 1}$$

confidence limits for the sample correlation coefficient can be evaluated. Some typical results are shown in Figure 5.3.

The 90% confidence limits are shown in the figure. Two trends can be seen: the confidence limits are virtually constant for more than about 30 data points and the width of the confidence region decreases as the correlation coefficient becomes larger. Consequently, small sample correlation coefficients should be regarded as effectively equal to zero; other values should be regarded as having modest precision.

Although the estimation of experimental errors can be quite tedious it should be seen as an essential part of the control of data quality—no set of data is complete without the error estimates. Subsequent analysis of the data using the least-squares method cannot be performed with confidence without realistic error estimates.

5.3.1 CAT VERSUS DOG

The process of obtaining data by coaddition was called CAT scanning in the 1970s (CAT stands for computation of average transients). The essence of CAT scanning is the coaddition, in order to reduce noise, of many data sets obtained quickly. However, there is another way in which noise can be reduced in some circumstances, usually by spending more time on each measurement. The circumstances are as follows. When the signal from the instrument is a time average of 'instantaneous' signals, the noise on that signal will be

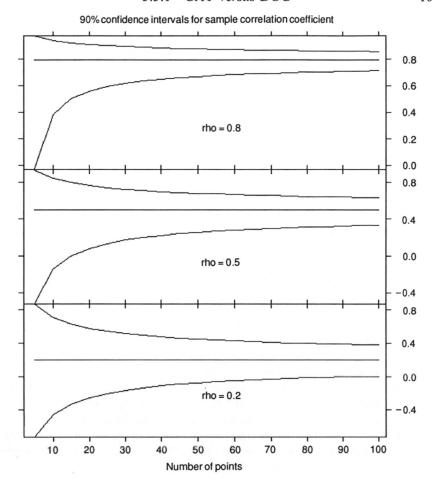

Figure 5.3. 90% confidence limits for sample correlation coefficient, rho, as a function of the number of measurements in the sample

inversely proportional to the square root of the measurement time. Taking more time may be termed DOG scanning; fewer sets of data are collected, but each set is of a higher quality as a result of being collected more carefully. To find the right balance between CAT and DOG one needs to be able to assess the relative effect on noise of the two processes. The former is easy, as it depends on the number of scans. The latter depends on the instrument characteristics, and again the information required is often not supplied by the instrument manufacturers. If that is the case it can be estimated by obtaining estimates of the errors, using the procedures given above, for the two lots of instrumental settings.

5.3.2 CAT OR DOG VERSUS SMOOTHING

The impatient experimenter may think, why spend time to reduce experimental noise when I can reduce it by polynomial smoothing (Section 7.4)? The answer is twofold.

On the one hand, slower data collection, where possible, improves the overall quality of the data by allowing more time for the measurement of each data point. It may help to reduce correlation between adjacent data points by leaving more time for the instrument to settle down between readings. It may be possible to reduce instrumental distortion of the data. Instrumental distortion arises because the signal produced by an instrument is the convolution of the true signal and an instrument transfer function. The ideal transfer function is a spike, or δ function, but real transfer functions are not usually so simple. By slowing down the scan rate and adjusting the instrument settings the transfer function can often be made more spike-like.

On the other hand, the noise reduction factor can be estimated for smoothing and for coaddition. For an m-point quadratic/cubic smooth the factor is approximately $2\sqrt{m}/3$ (see Section 7.4.3), whilst for m coadditions it is $\sqrt{m}$. How these factors compare is indicated in Figure 5.4. A given noise improvement factor may be achieved by coaddition of m scans or by smoothing with a $9m/4$-point quadratic/cubic smoothing function. Thus, noise may be reduced by a factor of 4 either by coaddition of 16 scans or by smoothing with a 36-point quadratic/cubic function. If the distortion caused by smoothing can be tolerated the best result would be obtained by doing both, which would result in a noise improvement factor of 16! It is clear that large reductions in noise, such as the factor of 100 mentioned above, can only be achieved practically by coaddition. However, the modest improvements in the signal-to-noise ratio afforded by smoothing can be used to great effect, as for example in lowering the detection limits in chemical analysis.

5.4 OUTLIERS

Outliers are data points which appear to belong to an alternate data set, i.e. they are data points subject to some kind of error not normally associated with the data being processed. In the context of a least-squares analysis, outliers give rise to very large residuals. The problem is this. How may one decide if a datum point is an outlier, and having identified an outlier, what should be done about it: should it be eliminated or treated in some other way?

How to treat outliers is a problem of very long standing (see Barnett 1978, 1983). The problem is that they represent an extreme form of experimental error, and as such have a finite probability of occurrence anyway. To eliminate a data point because it gives a large residual is to get a better fit to the other

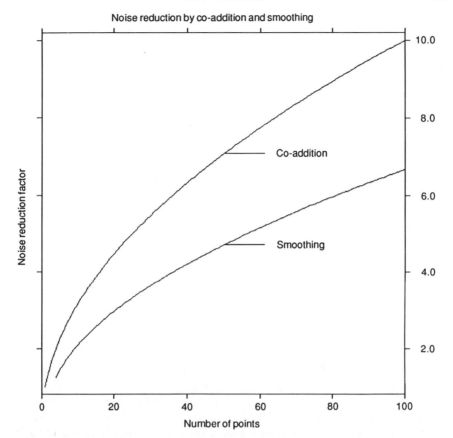

Figure 5.4. A comparison of the relative efficiencies of coaddition and smoothing for noise reduction. Upper curve: noise reduction factor for averaging m measurements. Lower curve: noise reduction factor for smoothing by an m-point quadratic/cubic polynomial convolution

points, but at the expense of reducing the errors on the parameters to unrealistically low values. On the other hand, inclusion of a rogue data point will also cause the parameter values and errors to be falsified.

Underlying this discussion there is an implicit idea that needs to be brought out into the open. The concept of an outlier only has meaning in the context of a model. For example, if we plot a set of data and see that all but one points lie approximately on a straight line, with the one point a long way off, we are implicitly making two assumptions: that the data should be represented by a linear model and that the outlier is an unexpectedly long distance from the line. 'Unexpectedly' is the key word. It implies that we think the error is much larger than it should be. Unfortunately our feel for what the error should be is conditioned by the errors (or residuals) observed for the other data points.

It is probably best to adopt a conservative approach to outliers. Firstly, one should check that there is no obvious unusual source of error, e.g. a blunder in copying out the data. Next the experimental procedure should be controlled to see if there is anything to which the data might be exceptionally sensitive. For example, temperature needs to be regulated in chemical kinetics and temperature control may not be adequate. One subject area that seems to be particularly prone to outliers is enzyme kinetics, presumably because of the sensitivity of enzymes to a number of external factors. Knowledge of the factors that may give rise to outliers should always be used to improve the experimental design and quality of the data. It is better by far to reduce the frequency of occurrence of outliers than to try to deal with them numerically.

When outliers cannot be eliminated experimentally, the data should be fitted both without them and with them. It may well be that the results are not significantly different. If that is so, there are no outliers and the problem disappears! If the results are significantly different, there is an intermediate stage between simply including or excluding the suspect data. This is to give those data points reduced weight on some kind of subjective basis. This is equivalent to the (subjective) assertion that the model is correct, but that the data need to be adjusted.

Even when data points are eliminated from a calculation, it is bad practice to eliminate them from the data set. A useful procedure is to set a flag for *all* data points such that when the flag is *up* the data point is included, and when it is *down* the point is excluded from the calculation. A third flag position may be used to indicate that a reduced weight should be used. In this way various calculations may be performed with selections from the complete data set.

6

Criteria for Model Selection

6.1 PROBABILITY AND EXPERIMENTAL ERROR

In this chapter we address the problem of deciding when a least-squares model should be regarded as acceptable or not. We shall begin by exploring the statistics of experimental error and the implications for residuals and parameters. This is followed by a section on hypothesis testing, with the emphasis on choosing the correct procedures and understanding their validity and limitations. Finally, we look at how the statistical results may be used both formally and informally to help with model selection.

When we make a measurement we record a number from the scale of the measuring device. This number is the quantity being measured divided by its physical dimensions. Experimental records are subject to errors, as discussed in Chapter 2. Let us now consider in more detail the nature of the random errors.

In the absence of systematic error, each observation can be considered as the sum of a 'true' value and an error:

$$y_i^{\text{obs}} = y_i^{\text{true}} + e_i$$

The 'true' value is a constant, whilst the error shall be considered as a *random* variable. The essence of a random variable is that the occurrence of a particular value cannot be predicted, that is, e_i cannot be predicted. However, if the experiment were to be repeated a large number of times we might be able to assign a probability that the error will take a particular value. To do this we need to know the probability distribution function.

As an introduction to the idea of a probability distribution function let us examine not the value, e_i, of the error, but another random variable, sign_i, which is given the value $+1$ when e_i is greater than or equal to zero, and a value of 0 when e_i is less than zero. (We could equally well call the values 'heads' and 'tails', making the variable exactly analogous to the outcome of tossing a coin.) We will make the assumption, which is intuitively very reasonable, that the values 0 and 1 are equally likely to occur.

Consider now what results we might expect if the measurement is repeated three times. There are eight possible outcomes, 000, 001, 010, 011, 100, 101, 110 and 111. Each of these outcomes is equally likely if we assume 0 and 1 are equally likely. The total probability is defined as unity. Therefore each of the eight possible outcomes is assigned a probability 1/8. From the individual probabilities we can derive probabilities for such outcomes as 'all results the same' (2/8), 'two results the same' (6/8) and so on, by counting the number of outcomes in the desired class.

Three zeros occurs once, two zeros three times, one zero three times and no zeros once. The *frequency* of occurrence of these classes of outcome is 1, 3, 3, 1, the coefficients of x in the binomial expansion $(1 + x)^3$. For this reason we say that our variable, sign$_i$, belongs to a *binomial* probability distribution. The general formula for the frequency of a particular outcome, x, in n repeated observations is given by

$$f(x) = \frac{n!}{x!(n - x)!} \, p^x(1 - p)^{n-x}$$

where x is the number of ones in the outcome and p is the probability that a single measurement will give an error whose sign value is 1. The assumption $p = \frac{1}{2}$ is not necessary in general. The mean and variance of the binomial distribution are np and $np(1 - p)$ respectively.

The binomial distribution is of limited value to us as we are more interested in the errors themselves, rather than their signs. The two most commonly used distribution functions are the Poisson and normal distributions.

The Poisson distribution can be derived from the binomial by assuming that n is large and p is very small. In the limit, assuming that np is a constant, λ,

$$f(x) = \frac{\lambda^x \, e^{-\lambda}}{x!}$$

In this case x is an integer that may take the values $0, 1, 2, ..., n$. The mean and variance of the Poisson distribution are both equal to the constant λ. The Poisson distribution is particularly applicable to the measurement of radioactive decay counts. A single radioactive nucleus may or may not decay in a given period of time, but in a certain mass of the substance, a fixed proportion of the nuclei present will decay in the given time period. Therefore, although the overall rate of decay is constant, i.e. the product np is constant, the actual number of decays observed in a given time interval is variable and can be interpreted in terms of the Poisson distribution.

The normal distribution has the following form

$$f(x) = \frac{1}{\sigma(2\pi)^{1/2}} \exp\left(\frac{-(x-\mu)^2}{2\sigma^2}\right) dx$$

This distribution differs from those we have discussed so far in that the variable x is continuous rather than discrete. For this reason the function gives the probability for the variable lying between the values x and $x + dx$; x may take any real value between plus and minus infinity. The mean of the normal distribution is μ and the variance is σ^2. If we make a change of variable

$$z = \frac{x - \mu}{\sigma}$$

we obtain the standard normal distribution

$$f(z) = \frac{1}{(2\pi)^{1/2}} e^{-z^2/2} dz$$

which has a mean of zero and a variance of one.

The binomial distribution, with the change of variable $z = (x - np)/[np(1-p)]^{1/2}$, tends towards the standard normal distribution in the limit that n goes to infinity. In practice the binomial distribution is well approximated by a normal distribution when np and $n - np$ are greater than 5. This is illustrated in Figure 6.1, which shows the binomial distribution for $n = 16$ and $p = 0.5$, together with the normal distribution with the same mean and standard deviation.

Also, the Poisson distribution, with the change of variable $z = (x - \lambda)/\lambda^{1/2}$, tends towards the standard normal distribution in the limit $\lambda = \infty$. It is approximated by the normal distribution when λ is larger than ca. 50. Figure 6.2 shows the Poisson distribution for $\lambda = 9$, together with the normal distribution having the same mean and standard deviation. The approximation is not as good as in the binomial case since the Poisson distribution is not symmetric about the mean—its moment coefficient of skewness is $1/\sqrt{\lambda}$.

The fact that the normal distribution is a good approximation to the binomial and Poisson distributions helps to explain why most statistics are based on the normal distribution; the errors introduced by using the wrong probability distribution are hopefully not large. Another reason lies in the so-called central limit theorem, which states that the sum of a number of random variables of given mean and variance has a normal probability distribution, regardless of the distributions of the individual variables. Therefore, if experimental error is to be regarded as the sum of a number of contributing errors of similar magnitude, it is not unreasonable that it should be assumed to have a normal probability distribution. Finally, the normal distribution is employed because it is relatively easy to handle.

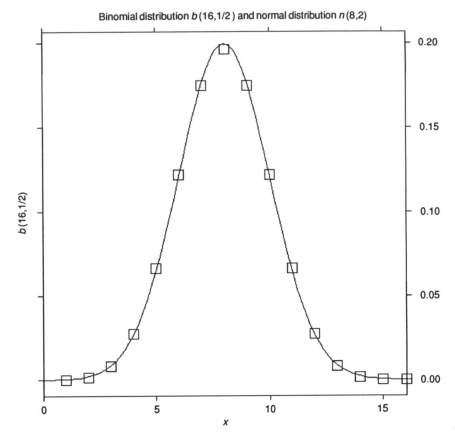

Figure 6.1. The binomial distribution $b(16, 1/2)$ (squares) and the normal distribution (line) of equal mean and standard deviation

By definition each error, e_i, has a mean value of zero, as the mean of a large number of observations should tend towards the 'true' value. It is usual to assume that each error is a sample from a population that has a normal distribution, except in cases such as radioactive decay, when the distribution is taken as Poisson. Although the assumption of a normal distribution is rarely based on sound theory, experience has shown that it is not a bad assumption. However, one should always bear in mind that it is an assumption which may, if one is unlucky, be a bad one.

Suppose now that a number of observations are made for different values of an independent variable. If each observation is divided by an estimate of its standard deviation, then the errors, e_i $(i = 1, m)$, may belong to a standard normal distribution of mean zero and unit variance. This assumes that there is no correlation of errors.

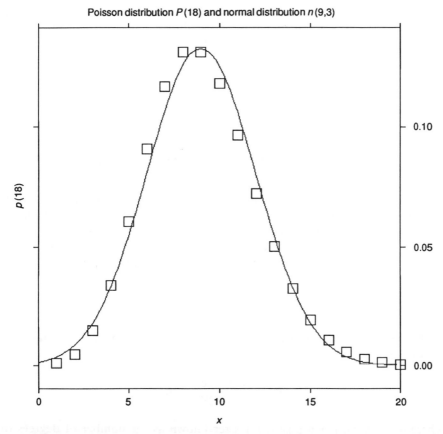

Poisson distribution $P(18)$ and normal distribution $n(9,3)$

Figure 6.2. The Poisson distribution $P(9)$ (squares) and the normal distribution (line) of equal mean and standard deviation

In the case that there is correlation we must describe the probability distribution of the errors as multivariate. The most frequently used multivariate distribution is the multivariate normal distribution. For m correlated random variables, $\mathbf{x}$, with a vector of means $\boldsymbol{\mu}$ and a variance–covariance matrix $\mathbf{M}$, it has the form

$$f(x) = \frac{1}{|2\pi\mathbf{M}|^{1/2}} \exp\left[\frac{-(\mathbf{x}-\boldsymbol{\mu})^{\mathrm{T}}\mathbf{M}^{-1}(\mathbf{x}-\boldsymbol{\mu})}{2}\right] \, \mathrm{d}x$$

This tends towards a set of standard normal distributions as the correlation coefficients tend towards zero, and $\mathbf{M}$ tends to a diagonal matrix of individual variances. Many of the results quoted in the following sections are valid for both univariate and multivariate probability distributions. Note particularly that the least-squares estimate of $\mathbf{r}^{\mathrm{T}}\mathbf{Wr}$ corresponds to the maximum likelihood value when the observations belong to a multivariate normal distribution, or to a set of univariate normal distributions if and only if the weight matrix $\mathbf{W}$ is equal to the inverse of the variance–covariance matrix of the observations. Also the scaled errors $M^{-1/2}e$ belong to a standard multivariate distribution with means of zero and a unit variance–covariance matrix. For more details of multivariate distributions see, for example, Mardia, Kent and Bibby (1979).

6.2 t, χ^2, F AND OTHER DISTRIBUTIONS

In this section we look at the probability distributions of functions of one or more random variables, but restrict our attention to the case where the latter belong to normal distributions.

If any variable X belongs to a normal distribution with mean zero and variance σ^2 then ΣX^2 belongs to the general χ^2 distribution:

$$f(x) = \frac{x^{r/2-1} \, e^{-x/2\sigma^2}}{\Gamma(r/2)2^{r/2} \, \sigma^r} \, dx$$

It follows that the scaled sum $\Sigma X^2/\sigma^2$ belongs to the standard χ^2 distribution:

$$f(x) = \frac{x^{r/2-1} \, e^{-x/2}}{\Gamma(r/2)2^{r/2}} \, dx$$

where $0 < x < \infty$, r is a positive integer known as the number of degrees of freedom and $\Gamma(z)$ is the so-called gamma function, which is a generalization of the factorial function. We may write the value for a given probability as $\chi^2(p, r)$. Note that the χ^2 distribution is a special case of the *gamma distribution*.[*] χ^2 is an unsymmetrical function with mean r, variance $2r$ and a maximum at $r-2$.

We next consider the distribution of such quantities as

$$T = \frac{X}{\sqrt{(Z/r)}}$$

If X is normally distributed with zero mean and unit standard deviation, and Z is distributed as χ^2 with r degrees of freedom, the variable T is said to belong

---

[*] An instance of the gamma distribution is as follows. $\Gamma(n)$ is the distribution of the waiting time for the nth event in a Poisson distribution of unit mean. $\Gamma(1)$ is the so-called *exponential* distribution.

to a *t* distribution with *r* degrees of freedom. The form of the *t*-distribution function is

$$f(x) = \frac{\Gamma[(r+1)/2]}{\sqrt{(\pi r)}\Gamma(r/2)(1+x^2/r)^{(r+1)/2}}\, \mathrm{d}x, \qquad -\infty < x < \infty$$

It is convenient to write the value of this function for a given probability as $t(p, r)$. The *t* distribution is symmetrical about its mean value of zero, and has a variance of $r/(r-2)$. As *r* increases *t* tends towards the standard normal distribution of the scaled variable $x(r-2)/r$. The approximation is moderately good for $r > 10$ and good for $r > 30$.

The ratio of two quantities that both belong to χ^2 distributions, typically the sums of squared residuals for two distinct least-squares models, divided by their respective number of degrees of freedom belongs to an *F* distribution:

$$F = \frac{U/r_1}{V/r_2}$$

The *F* distribution is said to have r_1 degrees of freedom in the numerator and r_2 degrees of freedom in the denominator, and has the standard form

$$f(x) = \frac{\Gamma[(r_1+r_2)/2]}{\Gamma(r_1/2)\Gamma(r_2/2)}\, r_1^{r_1/2} r_2^{r_2/2} x^{r_1/2-1} (r_2 + r_1 x)^{-(r_1+r_2)/2}$$

It is convenient to write the value of the *F* function as $F(p, r_1, r_2)$, where *p* is a probability. With this convention we can write down some useful relationships:

$$F(1-p, r_1, r_2) = 1/F(p, r_1, r_2)$$
$$F(1-p, 1, r) = (t_{1-p/2,r})^2$$
$$F(p, r, \infty) = \chi^2(p, r)/r$$

The latter two illustrate the relationship between the *F* distribution, the *t* distribution and the χ^2 distribution.

The *uniform* distribution is of interest because most random number generators in computer software generate numbers that belong to this distribution. Such random numbers may be used to simulate experimental error. The uniform distribution is defined as

$$f(x) = 1/(b-a), \qquad a \leqslant x \leqslant b$$

and has a mean of $(a+b)/2$ and a variance of $(b-a)^2/12$. Thus, a uniform distribution in the range $0 \leqslant x \leqslant 1$ has a mean of $\frac{1}{2}$ and a variance of $\frac{1}{12}$. Algorithms for portable random number generators are given in *Numerical Recipes* (Press *et al.*, 1986).

Experimental errors do not usually belong to a uniform distribution, but numbers belonging to one can be transformed into a set belonging to another distribution. General methods for doing this are given in *Numerical Recipes.* Here we will give only a method for generating a normal distribution. Two values, x_1 and x_2, from a uniform distribution on the interval $0 \leqslant x \leqslant 1$ are required:

$$z = \sqrt{(-2 \log_e x_1)} \times \cos 2\pi x_2$$

The transformed variable z belongs to a standard normal distribution of mean zero and unit variance.

6.3   EVALUATION OF DISTRIBUTION FUNCTIONS

The value of any distribution function or any integral of such a function can be calculated by means of a computer program. Mardia and Zemroch (1978) have published a complete set of algorithms for the normal, t, F, β and χ^2 distributions. The programs include both the evaluation of the function for a given probability, and the inverse calculation of the probability for a given value of the function. Another set of algorithms can be found in *Numerical Recipes* (Press *et al.*, 1986); these do not include the inverse calculations, but they do include the cumulative Poisson and binomial distribution functions. It is therefore possible to incorporate routines suitable for statistical testing into other computer programs, so that statistical testing may become part of the main, scientific program.

   The alternative and more traditional approach is to obtain the required information from published tables of the values. This procedure suffers from two limitations. Obviously it is necessary to have copies of the tables to hand and, secondly, the tables may not contain the values required for a specific test, so that the user must perform some kind of interpolation. Clearly, computer programs that can use any input are preferable to the use of tables.

   In either case the user should be aware that the tables or programs are very specific and not always the same. For instance, the cumulative normal distribution may take either of the following forms:

$$f(x) = \int_{-\infty}^{x} \frac{1}{(2\pi)^{1/2}} e^{-z^2/2} \, dz$$

or

$$g(x) = \int_{0}^{x} \frac{1}{(2\pi)^{1/2}} e^{-z^2/2} \, dz$$

For $x > 0$ these forms are related by $g(x) = f(x) + 0.5$. It is important to know which form of calculation has been used to produce the values.

   Another trap for the unwary is that some tables of the t function are given in terms of the confidence level, whilst others are given in terms of the significance level.

6.4

6.4.1   PROBABILITY AND RESIDUALS

In a least-squares analysis we minimize the weighted sum of squared residuals, so we must ask: what is the statistical relation between residuals and experimental error? The following theory applies to linear least-squares models of the form

$$\mathbf{y}^{obs} = \mathbf{y}^{calc} + \mathbf{e}$$
$$\mathbf{y}^{calc} = \mathbf{Jp}$$

where $\mathbf{e}$ is the vector of experimental errors. The theory is approximate for non-linear models.

Let us suppose that the variance–covariance matrix of the observations is $\mathbf{M}$ and has been factorized, $\mathbf{M} = \mathbf{m}^T\mathbf{m}$. Let us take the weight matrix to be the inverse of $\mathbf{M}$, $\mathbf{WM} = \mathbf{I}$:

$$\mathbf{M}^{-1} = \mathbf{m}^{-1}(\mathbf{m}^T)^{-1} = \mathbf{w}^T\mathbf{w} = \mathbf{W}; \qquad \mathbf{mw}^T = \mathbf{I}$$

(In the common case that $\mathbf{W}$ is diagonal, $w_{ii}^2 = W_{ii}$). If we define the weighted observations, Jacobian and residuals as

$$\hat{\mathbf{y}} = \mathbf{wy}^{obs}; \qquad \hat{\mathbf{J}} = \mathbf{wJ}; \qquad \hat{\mathbf{r}} = \mathbf{wr}$$

then the variance–covariance matrix of the weighted observations is given by the law of propagation of error (Section 2.2.1) to be a unit matrix:

$$\mathbf{M}(\hat{\mathbf{y}}) = \mathbf{wMw}^T = \mathbf{I}$$

Since the residual vector is defined as $\mathbf{r} = \mathbf{y}^{obs} - \mathbf{Jp}$, the weighted residual vector is given by

$$\mathbf{wr} = \mathbf{wy}^{obs} - \mathbf{wJp}; \qquad \hat{\mathbf{r}} = \hat{\mathbf{y}} - \hat{\mathbf{J}}\mathbf{p}$$

and since $\mathbf{p} = (\mathbf{J}^T\mathbf{WJ})^{-1}\mathbf{J}^T\mathbf{Wy}^{obs}$ and so $\mathbf{p} = (\hat{\mathbf{J}}^T\hat{\mathbf{J}})^{-1}\hat{\mathbf{J}}^T\hat{\mathbf{y}}$, then

$$\hat{\mathbf{r}} = \hat{\mathbf{y}} - \hat{\mathbf{J}}(\hat{\mathbf{J}}^T\hat{\mathbf{J}})^{-1}\hat{\mathbf{J}}^T\hat{\mathbf{y}}$$

The matrix $\hat{\mathbf{J}}(\hat{\mathbf{J}}^T\hat{\mathbf{J}})^{-1}\hat{\mathbf{J}}^T$ is called the hat matrix in the statistical literature, and is given the symbol $\mathbf{H}$:[*]

$$\mathbf{H} = \hat{\mathbf{J}}(\hat{\mathbf{J}}^T\hat{\mathbf{J}})^{-1}\hat{\mathbf{J}}^T$$
$$\hat{\mathbf{r}} = (\mathbf{I} - \mathbf{H})\hat{\mathbf{y}}$$

---

[*] The hat matrix, $\mathbf{H}$, is said to be idempotent, $\mathbf{HH} = \mathbf{H}$. Properties that follow from this include: $H_{ii} = \Sigma H_{ik}^2$, $0 \leqslant H_{ii} \leqslant 1$, it has n unit eigenvalues and $m - n$ zero eigenvalues; $\Sigma H_{ii} = n$. $\mathbf{I} - \mathbf{H}$ is also idempotent, $(\mathbf{I} - \mathbf{H})(\mathbf{I} - \mathbf{H}) = \mathbf{I} - \mathbf{H}$.

The expression $\hat{\mathbf{r}}^T\hat{\mathbf{r}} = \hat{\mathbf{y}}^T(\mathbf{I} - \mathbf{H})\hat{\mathbf{y}}$ provides a means of calculating the objective function without calculating the residuals. Use of this facility is to be discouraged unless the residuals are also calculated separately.

The sum of squared weighted residuals is given by

$$\hat{r}^T\hat{r} = \hat{y}^T(I - H)(I - H)\hat{y}$$
$$= \hat{y}^T(I - H)\hat{y}$$
$$= \hat{y}^T\hat{y} - \hat{y}^TH\hat{y}$$
$$= (y^{obs})^TWy^{obs} - p^T(J^TWJ)p$$

It can be shown (see Appendix 5 for the proof), and is a very important result, that this sum of squares has an expectation value of $m - n$.

If the variance–covariance matrix is equal to $M\sigma^2$, where σ^2 is the unknown variance of an observation of unit weight, and $W = M^{-1}$, then it follows that the expectation value of S is

$$E(S) = E(\hat{r}^T\hat{r}) = (m - n)\sigma^2$$

and an estimate of σ^2 is given by $S/(m - n)$. This commonly comes into play when the weights are assumed to be equal and are assigned a unit matrix.

These results are independent of the nature of the distribution function of the experimental errors. The only assumptions are that the errors have zero means and finite variances. If, however, we are prepared to make assertions concerning the distribution function of the errors we may derive something of the distribution function of the residuals. The most common assumption is that the errors belong to a multivariate normal distribution with mean vector zero and variance–covariance matrix M. It is convenient to use this assumption even when the observations are not correlated (M is diagonal) as the general results are always applicable.

From the equality $\hat{r} = (I - H)\hat{y}$ we can see that the residuals are linear combinations of the observations. As a consequence of this we may use a theorem from mathematical statistics to derive the probability distribution of the residuals. The theorem is (Mardia, Kent and Bibby, 1979, Theorem 6.2.2) that if the observation errors belong to a multivariate normal distribution, then so do the residuals.

The means (expectation values) of the residuals are zero (Mardia, Kent and Bibby, 1979, Theorem 6.2.2). In the common case that the least-squares model contains a constant parameter (such that the Jacobian has a column of ones) the actual weighted mean of all the residuals is also zero. This is proved by multiplying the residual vector on the left by J^TW:

$$J^TWr = J^TWy^{obs} - J^TWJ(J^TWJ)^{-1}J^TWy^{obs} = 0$$

J^TWr is simply the weighted sum of residuals when a row of J^T consists of ones.

When the errors belong to a multivariate normal distribution with mean vector zero and variance–covariance matrix M, and $W = M^{-1}$, the sum of weighted squared residuals, $\mathbf{r}^{\mathrm{T}}\mathbf{W}\mathbf{r}$, is distributed as χ^2 with $m - n$ degrees of freedom. Hence, $\mathbf{r}^{\mathrm{T}}\mathbf{W}\mathbf{r}/(m-n)$ is distributed as $\chi^2/(m-n)$ with $m - n$ degrees of freedom; we may say it belongs to a reduced χ^2 distribution.

The variance–covariance matrix for the residuals is given by

$$\mathbf{M}(\hat{\mathbf{r}}) = (\mathbf{I} - \mathbf{H})(\mathbf{w}\mathbf{M}\mathbf{w}^{\mathrm{T}})(\mathbf{I} - \mathbf{H})^{\mathrm{T}}$$
$$= (\mathbf{I} - \mathbf{H})\mathbf{I}(\mathbf{I} - \mathbf{H})^{\mathrm{T}}$$
$$= \mathbf{I} - \mathbf{H}$$

The variance of any one residual is simply given by the diagonal elements of $\mathbf{I} - \mathbf{H}$:

$$s^2(\hat{r}_i) = (1 - H_{ii})\sigma^2$$

When the weights have been correctly assigned, σ^2, the variance of an observation of unit weight, is unity. It follows that each residual, when correctly weighted and divided by the appropriate scaling factor, belongs to a standard normal distribution of mean zero and unit variance:

$$\frac{r_i - \mu(r_i)}{\sqrt{(1 - H_{ii})}} \sim n(0, 1)$$

It is not common practice to calculate the hat matrix so this scaling is not in common use, but this is a shame as residuals scaled in this way belong to a single probability distribution and so can be treated in a standard way.

If the weights are only proportional to the inverse of the variance–covariance matrix of the observations, then σ^2 will have to be estimated as $\mathbf{r}^{\mathrm{T}}\mathbf{W}\mathbf{r}/(m-n)$. In this case the scaled residuals

$$\frac{r_i - \mu(r_i)}{\sqrt{[\mathbf{r}^{\mathrm{T}}\mathbf{W}\mathbf{r}(1 - H_{ii})/(m-n)]}}$$

belong to a t distribution with $m - n$ degrees of freedom. If $m - n$ is large (greater than about 30) this approximates well to a standard normal distribution. Residuals in this form are referred to in the statistics literature as 'internally studentized'.

To show the value of the hat matrix, let us consider fitting a straight line $y = a + bx$ to seven equally spaced and equally weighted data points $(x = -3, -2, ..., 2, 3; W_{ii} = 1)$. The ith row of the Jacobian is $[1 \; x_i]$ so the diagonal elements of the hat matrix are given by $H_{ii} = (4 + x_i^2)/28$. Therefore the variances of the residuals are in the ratio $15:20:23:24:23:20:15$. This shows that, even when the observations may have equal variance, the residuals do not have equal variances, though in this instance the departure from constant variance is not large. The matrix of correlation coefficients for this example is

$$\rho(r_i, r_j) = \begin{bmatrix} 1.00 \\ -0.58 & 1.00 & & & & & \text{symmetrical} \\ -0.38 & -0.28 & 1.00 \\ -0.21 & -0.18 & -0.17 & 1.00 \\ -0.05 & -0.09 & -0.13 & -0.17 & 1.00 \\ 0.12 & 0.00 & -0.09 & -0.18 & -0.28 & 1.00 \\ 0.33 & 0.12 & -0.05 & -0.21 & -0.38 & -0.58 & 1.00 \end{bmatrix}$$

Note that the residuals are correlated even when the observations are not.

6.4.2 PROBABILITY AND CALCULATED VALUES

Using the same formalism as in the previous section, the weighted calculated values can be seen to be linear combinations of the observations

$$\mathbf{w}\mathbf{y}^{\text{calc}} = \mathbf{H}\hat{\mathbf{y}}$$

where $\mathbf{H}$ is the hat matrix. It follows that the weighted calculated values are very likely to be normally distributed. The expectation values are, of course, $\mathbf{w}\mathbf{y}^{\text{true}}$, and the variance of the ith calculated value is simply H_{ii} when σ^2 is known and $\mathbf{r}^T\mathbf{W}\mathbf{r}H_{ii}/(m-n)$ when it is not. It is interesting to note that the ratio of the variance of a residual to the variance of the calculated value is $(1 - H_{ii})/H_{ii}$. Furthermore, the variance of a calculated value is always less than the variance of the corresponding observation because H_{ii} is less than unity. This point is taken up quantitatively in Section 7.4.3.

A knowledge of the variances of the calculated values can be used to set confidence limits for the calculated curve. Mandel (1964) and others give examples. There is also correlation between calculated values. The variance–covariance matrix is equal to $\mathbf{H}\mathbf{M}(\hat{\mathbf{y}})\mathbf{H}^T$, so that in the case of observations of equal variance the correlation coefficients are given by $H_{ij}/(H_{ii}H_{jj})^{1/2}$. The question of correlation between calculated values is taken up again in Section 7.4.6.

6.5 PROBABILITY AND PARAMETERS

As with residuals, the least-squares parameters are linear combinations of the observations

$$\mathbf{p} = (\mathbf{J}^T\mathbf{W}\mathbf{J})^{-1}\mathbf{J}^T\mathbf{W}\mathbf{y}^{\text{obs}}$$

Therefore, if the errors are normally distributed, the variance of an observation of unit weight is known and $\mathbf{W} = \mathbf{M}(\mathbf{y})^{-1}$; the parameters belong to normal distributions with means given by $(\mathbf{J}^T\mathbf{W}\mathbf{J})^{-1}\mathbf{J}^T\mathbf{W}\mathbf{y}^{\text{true}}$ and variances given by the diagonal elements of $[(\mathbf{J}^T\mathbf{W}\mathbf{J})^{-1}\mathbf{J}^T\mathbf{W}]\mathbf{M}(\mathbf{y})\,[(\mathbf{J}^T\mathbf{W}\mathbf{J})^{-1}\mathbf{J}^T\mathbf{W}]^T$ or $(\mathbf{J}^T\mathbf{W}\mathbf{J})^{-1}$. With this assumption we are in a position to ask some pertinent questions regarding the parameters of a least-squares model.

If the calculation gives us a parameter value of p with a standard deviation of s_p, what is the probability that the parameter has a 'true' value lying between p and $p + s_p$? The answer is given as the area under the probability function between $x = p$ and $x = p + s_p$:

$$F(x) = \int_p^{p+s_p} \frac{1}{s_p(2\pi)^{1/2}} \exp\left(\frac{-(x-p)^2}{2s_p^2}\right) \mathrm{d}x$$

$$= \int_0^1 \frac{1}{(2\pi)^{1/2}} \mathrm{e}^{-z^2/2}\,\mathrm{d}z, \qquad z = (x-p)/s_p$$

This is effectively the sum of all the probabilities that p takes a certain value in the stated interval. The integral may be evaluated either by a computer program or by consulting statistical tables. Its value is 0.3413. In a similar way the probability that the parameter has a 'true' value lying between p and $p - s_p$ is also 0.3413. Therefore the probability that the parameter has a 'true' value lying between $p - s_p$ and $p + s_p$ is 0.6813. In other words, there is a 68% chance that the parameter will lie within one standard deviation of its calculated value and a 32% chance that it will lie outside this region. The chance of a parameter being more than two standard deviations away from its calculated value is 4.6%, while the chance of a parameter being more than three standard deviations away from its calculated value is less than 1%. A cautious scientist would therefore expect the parameter to be within two or three standard deviations of the value given by the least-squares calculation.

This example shows how, assuming a given probability distribution function for the errors, we may calculate probabilities of direct relevance to scientific investigations. However, the choice of the probability level at which a result is acceptable is a subjective one. For instance, if we wish there to be exactly a 1% probability that a parameter lies outside a range, we would select the range as $\pm 2.57 s_p$.

The two figures, 68 and 32%, are known respectively as the *confidence* and *significance* levels. The 68% confidence region, also known as the critical region, is the region $p \pm s_p$. Thus, 68% is the probability of finding a value inside the confidence limits $p \pm s_p$ and 32% is the probability of finding a value outside the confidence limits.

It is a remarkable fact that an upper bound can be placed on the significance level regardless of any assumptions concerning probability distributions. According to Chebychev's inequality, which is valid for all random variables,

$$P(\,|X - \mu| \geq n\sigma) \leq 1/n^2$$

Thus, the maximum probabilities that a parameter will be more than 1, 2 or 3 standard deviations away from its expectation value are 100, 25 and 11%.

These bounds should be applied when the number of observations is 'small' in order not to overestimate the precision of the parameters.

So far we have assumed that the variance of an observation of unit weight is known. When it is not known it may be estimated as $S = \mathbf{r}^T \mathbf{W} \mathbf{r} / (m - n)$, and the variances on the parameters are given by $(\mathbf{J}^T \mathbf{W} \mathbf{J})_{pp}^{-1} S$. In this case the function $| p_i - p^{\text{test}} | / \sigma_p$ is distributed as $t(\alpha + \frac{1}{2}, m - n)$, where p^{test} is a value of the parameter to be used in hypothesis testing.

To make a simultaneous test on two or more parameters is rather more complicated (Hamilton, 1964). We suppose that there is a linear relationship $\mathbf{Tp} = \mathbf{z}$. Then

$$\frac{(\mathbf{Z} - \mathbf{Tp})^T (\mathbf{T} \mathbf{M}_p \mathbf{T}^T)^{-1} (\mathbf{Z} - \mathbf{Tp})}{t}$$

is distributed as $F_{t, m-n}$. $\mathbf{M}_p$ is the relevant part of the variance–covariance matrix of parameters and t is the rank of the $\mathbf{T}$ matrix.

To illustrate this consider the following example with a three-parameter model. We wish to test the hypothesis

$$H_0 : p_1 = p_3$$

where $p_1 = p_3$ is equivalent to

$$[1 \quad -1] \begin{bmatrix} p_1 \\ p_3 \end{bmatrix} = 0$$

that is, $\mathbf{T} = (1 \ -1)$, $z = 0$. The rank of $\mathbf{T}$ is 1. For this case the function reduces to $(p_1 - p_3)^2 / (\sigma_1^2 - 2\rho_{13} + \sigma_3^2)$, which is tested against $F(\alpha, 1, m - n)$ or the equivalent, $t(\alpha + \frac{1}{2}, 1, m - n)^2$.

Another example could be taken from the coin mass data discussed in Chapter 1. Fitting a straight line to the mean masses for different years gives estimates of the intercept, p_1, and the slope, p_2. Suppose we wished to test the hypothesis

$$H_0 : p_1 = 3.564 \text{ and } p_2 = 0$$

The equalities are equivalent to

$$\begin{bmatrix} 1 & 0 \\ 0 & 1 \end{bmatrix} \begin{bmatrix} p_1 \\ p_2 \end{bmatrix} = \begin{bmatrix} 3.564 \\ 0 \end{bmatrix}$$

Since $\mathbf{T}$ is an identity matrix of rank 2, we must test

$$\frac{(Z - \mathbf{p})^T \mathbf{M}_p^{-1} (Z - \mathbf{p})}{2}$$

against $F(\alpha, 2, m - n)$. Writing $\mathbf{W}$ for $\mathbf{M}_p^{-1}$, this expression becomes

$$(3.564 - p_1)^2 W_{11} - 2(3.564 - p_1) p_2 W_{12} + p_2^2 W_{22}$$

To perform this test one needs only to substitute the actual values for the parameters and their variance–covariance matrix. It should be noted that when making tests on two or more parameters the correlation between parameters *must* be taken into account, as shown by the appearance of the term W_{12} above.

6.6

6.6.1   HYPOTHESIS TESTING

Hypothesis testing is a formal statistical procedure which provides rules for making decisions. The central concept of hypothesis testing is the *null hypothesis*, H_0. The null hypothesis is a statement of the form

$H_0: a$ is {relation} to b;   {relation} $= \{ = , \neq, >, <, \geqslant, \text{or} \leqslant \}$

The question we must ask is, should we accept or reject the null hypothesis? If we reject the null hypothesis we may make an alternative suggestion. Any alternative that we wish to test is designated as H_1.

Along with acceptance or rejection there are two types of error that we might make.* A type I error is made when we reject a null hypothesis that should be accepted. In hypothesis testing we must make a judgement of the risk we are willing to take of making a type I error. This is a subjective judgement that can only be made in the light of experience. The risk can be quantified as the probability of rejecting a true hypothesis. Expressed as a percentage it is known as the *level of significance*, α.

We have already discussed the concept of a significance level in terms of the normal distribution. However, it is instructive to look at it again from first principles in order to see how it depends on the sample probability distribution and the sample size. Consider an experiment designed to test the hypothesis that the probability of throwing heads when tossing a coin is $\frac{1}{2}$:

$$H_0: P(\text{heads}) = \tfrac{1}{2}$$

We toss the coin a number of times. The probable outcomes follow a binomial distribution, the probability of a given number of heads occurring for $p = \frac{1}{2}$ being given by Pascal's triangle (Table 6.1).

Let us now calculate the probability of making a type I error by rejecting the hypothesis when all the coins tossed are heads, or all are tails. Both the

---

* A type II error is made when we accept the null hypothesis when it should be rejected in favour of the alternative hypothesis. If the probability of H_1 being true is p, the *power* of an hypothesis test is said to be $1 - p$. When more than one hypothesis test can be made it is usual to employ the most powerful test.

Table 6.1. Probabilities of obtaining m heads in n throws of a fair coin. The probabilities have been multiplied by the total probability to give integers rather than fractions

Throws n	Number of heads m											Total
	0	1	2	3	4	5	6	7	8	9	10	
1	1	1										2
2	1	2	1									4
3	1	3	3	1								8
4	1	4	6	4	1							16
5	1	5	10	10	5	1	.					32
6	1	6	15	20	15	6	1					64
7	1	7	21	35	35	21	7	1				128
8	1	8	28	56	70	56	28	8	1			256
9	1	9	36	84	126	126	84	36	9	1		512
10	1	10	45	120	210	252	210	120	45	10	1	1024

all-heads outcome and the all-tails outcome are possible when the null hypothesis is true, and have a probability of 2^{-n} for n coin tosses. The probability of falsely rejecting the hypothesis is given by the probability of the unlikely events occurring. Clearly this decreases from $\frac{1}{2}$ for two coin tosses to $(1 + 1)/512$ for nine tosses. If we are prepared to take a 0.4% risk of making a type I error, nine coin tosses would be sufficient; we would only reject the hypothesis falsely if the outcome were to be nine heads or nine tails. If the risk we are prepared to take is 0.001, we would need to make 11 tosses, i.e. we would accept the hypothesis that the coin is fair at the 0.1% significance level if in 11 tosses we do not get either all heads or all tails.

This example illustrates how the probabilities are calculated by making an assumption concerning the nature of the probability distribution, and how the choice of significance level and the sample size are both relevant to the hypothesis test.

By considering both the all-heads and all-tails outcomes we have made a *two-tailed* test, because we have looked at both tails of the distribution. If we are considering the hypothesis

$$H_0: p > \tfrac{1}{2}$$

we should only consider the outcome of all heads. Generally speaking, we use a two-tailed test for an equality hypothesis and a one-tailed test for an hypothesis containing a comparison (greater or less than) unless the comparison is being made to an absolute value.* As an example let us find the

---

*A test of the type $H_0: |X - v| > 0$ is equivalent to the two conditions $X > v$ or $-X < -v$.

significance level for testing this hypothesis against the outcome that all, or all but one, of the coins tossed come out heads. For nine tosses the probability is $(9 + 1)/512$, or nearly 0.02. If we wish to test against the outcome two or less tails, the significance level would be $(37 + 9 + 1)/512$, or 0.09. In calculating this we are summing the probabilities of the individual outcomes. The sums, or integrals in the case of continuous probability distributions, are known as the *cumulative* probability distributions. Part of the cumulative binomial distribution is give in Table 6.2. The cumulative probability table is read as follows. The probability of getting m or less heads in n throws is given by the figure in the mth column and nth row. For example, the probability of getting 0, 1, 2 or 3 heads in seven throws is 0.5, and in eight throws is 93/256, or ca. 0.35. Notice that the cumulative probability of all possible outcomes is always unity.

To form the null hypothesis we must select a test statistic and relate it to a predetermined quantity. The hypothesis is tested by reference to values obtained from the cumulative probability distribution function. The significance level at which the hypothesis may be falsely rejected is found when the test statistic is equal to the calculated value. This is known as the critical value. If the test statistic is below the critical value the hypothesis may be accepted.

Consider the example of coin masses with the data given in Table 1.1. The published expectation value, or mean μ, for the mass is 3.564 g. We wish to ask if the sample average, m_0, is significantly different from the expectation value. Using the results of Section 6.5, we know that $| p_i - p^{\text{test}} |/s_p$ belongs to a t distribution with $m - n$ degrees of freedom. Therefore,

$$\frac{| m_0 - \mu |}{s_m}$$

belongs to a t distribution with $m - 1$ degrees of freedom. We may test

Table 6.2. Probabilities of obtaining m or less heads in n throws of a fair coin

Throws n	Number of heads									
	0	1	2	3	4	5	6	7	8	9
1	1/2	1								
2	1/4	1/2	1							
3	1/8	1/2	7/8	1						
4	1/16	5/16	11/16	15/16	1					
5	1/32	6/32	1/2	26/32	31/32	1				
6	1/64	7/64	22/64	42/64	57/64	63/64	1			
7	1/128	8/128	29/128	1/2	99/128	120/128	127/128	1		
8	1/256	9/256	37/256	93/256	163/256	219/256	247/256	255/256	1	
9	1/512	10/512	67/512	130/512	1/2	382/512	466/512	502/512	511/512	1

the hypothesis

$$H_0 : \mu = 3.564$$

with the values $m_0 = 3.5592$, $m = 31$, $s_m = 0.0284$; $T = 0.0048/0.0284 = 0.17$. This is less than $t(29)$ at the 80% significance level (two-tailed test), so we cannot reject the hypothesis at any reasonable significance level. This results from the fact that the difference from the test value is a small fraction of the standard deviation.

Suppose that the divergence had been two standard deviations. In that case we would accept the hypothesis at the 5% significance level, but reject it at the 4% significance level.

The quantity or $\mathbf{r}^T \mathbf{W} \mathbf{r} / (m - n)$ belongs to a reduced χ^2 distribution. In an ideal situation (Section 6.4) it should have the value unity. The reduced χ^2 is obtained from the normal χ^2 by dividing by $m - n$. We may therefore test the hypothesis

$$H_0 : \mathbf{r}^T \mathbf{W} \mathbf{r} / (m - n) = 1$$

by looking up the critical values of χ^2. For example, the hypothesis would be rejected at the 5% significance level if, for 7 degrees of freedom, $S > 2$. As the number of data points increases, the reduced χ^2 tends towards a constant value so for $m - n > 200$ the critical values of χ^2 are 1.17, 1.25 and 1.34 at the 5, 1 and 0.1% significance levels respectively.

These examples give some idea as to how hypotheses may be tested in the context of least-squares model testing. Note that these tests are based on the assumption that a parameter is normally distributed. For non-linear models there is an extra element of uncertainty, but one that is difficult to quantify.

6.6.2   CONFIDENCE LIMITS

An alternative to hypothesis testing is to establish confidence limits. Suppose we wished to establish the confidence limits for a least-squares parameter. This is equivalent to requiring that

$$- a(\alpha/2) < \frac{|p - p^{\mathrm{exp}}|}{s_p} < a(\alpha/2)$$

at a given confidence level. Here p is the value of the parameter and s_p is its standard deviation; p^{exp} is the expectation value of that parameter and is often denoted as μ, the mean, and $a(\alpha)$ is the value of a test statistic at the significance level α. Two cases must be distinguished: either the variance of an observation of unit weight is known or it is estimated as $\mathbf{r}^T \mathbf{W} \mathbf{r} / (m - n)$. In the former case the probability distribution function of the parameter determines which statistic shall be used. Thus, if the parameter p is assumed to belong to a normal distribution, $a(\alpha)$ is taken from a normal distribution. In this case

$a(0.05)$ is 1.65. There is a 95% chance that p^{exp} will be less than $p + 1.65s_p$. Likewise, there is a 95% chance that p^{exp} will be greater than $p - 1.65s_p$. Therefore there is a 90% chance that p^{exp} will lie in the region spanned by 1.65 standard deviations.

If the variance of an observation of unit weight is estimated, we must use a t test, though when the number of observations greatly exceeds the number of parameters $(m - n > \text{ca. } 30)$, $(m - n - 2)t(j, m - n)/(m - n)$ is well approximated by the standard normal distribution and the latter can be used directly with little error. Consider the mean of the coin masses (data in Table 1.1). There are 31 observations and 1 parameter, $m - n = 30$. The critical value of t for 30 degrees of freedom at the 5% significance level is 1.7. Therefore,

$$\frac{|m_0 - \mu|}{s_m} < 1.7$$

whence $m_0 - 1.7s_m$ is the lower confidence limit and $m_0 + 1.7s_m$ is the upper confidence limit. Therefore the 90% confidence limits are 3.51–3.61 g.

6.7

6.7.1   CRITERIA FOR ACCEPTING A MODEL

The question we shall address in this section is as follows. Is the least-squares model an adequate representation of the experimental data within the limits of experimental error? To answer this question we must look at such statistics as the sum of squared residuals, the standard deviations and correlation coefficients of the parameters, and the distribution of the residuals. These statistics are not mutually exclusive. On the contrary, a model should only be accepted if *all* the indicators are favourable.

6.7.1.1   The Sum of Squared Residuals

If the data have been given their correct weights the objective function, $\mathbf{r}^T\mathbf{W}\mathbf{r}$, has an expectation value equal to $m - n$, the number of data points less the number of parameters. The sum of squares is a difficult statistic to use by itself because departures from the expectation value may be due to a number of possible causes, but since the expectation value of $\mathbf{r}^T\mathbf{W}\mathbf{r}$ does not depend on the probability distribution of the experimental errors it is potentially the most reliable statistic to use for model selection.

Systematic errors in the data are one possible cause of deviations from the expectation value. For example, in potentiometric titration it is extremely difficult to eliminate systematic differences between duplicate titrations. One source of systematic error is to be found in the process of electrode calibration.

A small calibration error in each of two titration curves constitutes a systematic error which may be obscured when treating each curve individually by a small compensatory shift in parameter values; when treating the two curves together this compensation is not possible, and the systematic errors become manifest in a larger sum of squares. An incorrect model can also be regarded as introducing a systematic error. In particular a missing parameter will usually result in a large sum of squares.

The expectation value of $m - n$ is predicated on the weight matrix being the inverse of the correct variance–covariance matrix. Since the latter is very difficult to determine experimentally it is likely that incorrect weights will be an underlying cause of departure from the expectation value. Moreover, if it is assumed that the errors are uncorrelated when there is in fact correlation, i.e. if a diagonal weight matrix is used when the covariances are not indeed zero, the weight will always be incorrect.

Not weighting the data explicitly is equivalent to assigning unit weights to the residuals. Therefore in this case it is said that the sum of squares divided by $m - n$ is an estimate of the variance of an observation of unit weight, s^2. Since unit weights imply that all the observations have equal uncertainty, the square root, s, is an estimate of the average experimental error in an observation. To establish whether a model is adequate one needs to assess the average experimental error. If s is within a factor of three of this estimate the model may be acceptable.

From this discussion it is clear that the acceptance or rejection of a model depends on an assessment of experimental errors. Either the errors are translated into weights, in which case the objective function should have a value near to $m - n$, or the estimate of the variance of an observation of unit weight should be approximately equal to an estimate of the average experimental error.

6.7.1.2 Distribution of Residuals

This is as important as the sum of squares, if not more so. The first thing to do is to examine the residuals visually. This means plotting the residuals as vertical bars, as shown in Figure 6.3. The plot is best scaled to encompass the largest and smallest residual and is examined for the presence of systematic trends. In this figure there is a clear systematic trend, which was produced by trying to fit a straight line to data generated with a parabolic dependence on the independent variable.* This shows the value of plotting residuals when the observed and calculated data appear to be in good agreement with the

---

*These data were generated with the function

$$y = 1 + 2x + 0.01x^2 \qquad (x = 1, 2, ..., 10)$$

with no added noise. The straight line was fitted by the least-squares method with unit weights.

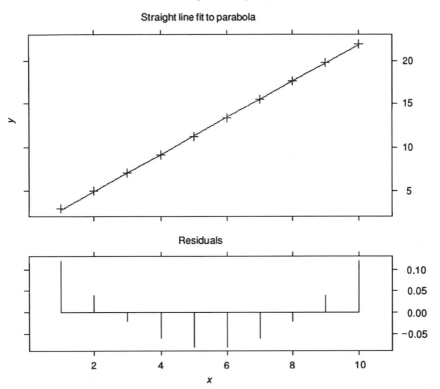

Figure 6.3. Upper box: best-fit straight line applied to data from a simulated noise-free parabola $y = 1 + 2x + 0.01x^2$. Lower box: residuals

observations. The residual plot may also indicate the presence of outliers (Section 5.4).

It is reasonable to expect that the weighted residuals, when scaled by the factor $(1 - h_{ii})^{-1/2}$, will belong to a standard normal distribution. One may confidently expect that there will be an approximately equal number of positive and negative weighted residuals (scaling with the hat matrix does not change the sign of any residual). Since the mean of all residuals is, in most cases, zero, the number of positive and negative residuals will be significantly different only if one class contains much bigger absolute values than the other. Such a situation should be obvious by inspection.

A criterion of historical interest is that, in a set of residuals obtained in sequence, there should not be long *runs* of residuals having the same sign. This is equivalent to saying that there should not be systematic trends in the sequence of residuals. Unfortunately it is difficult to quantify what constitutes a 'long' run.

The distribution of scaled weighted residuals can be tested by means of a χ^2 goodness-of-fit test; if the variance of an observation of unit weight is known then it is tested against a normal distribution, otherwise it is tested against a t distribution. We may divide the residuals into groups by magnitude and check to see how the number found in each group compared to the number expected, by consulting tables for the normal distribution. An easy way to do this is as follows. Say we choose to divide the residuals into eight groups, each expected to contain an equal number, an eighth of the total number. We look up the values that give cumulative probabilities of $\frac{1}{8}$, $\frac{2}{8}$, $\frac{3}{8}$, etc. These limits are then used to sort the residuals into groups. For the example of eight groups the limits, in terms of the normal distribution, are

$$-\infty \qquad -1.15 \qquad -0.675 \qquad -0.319 \qquad 0 \qquad 0.319 \qquad 0.675 \qquad 1.15 \qquad \infty$$

The expected population of each group is an eighth of the total number of residuals. For the jth group we calculate the actual population and a partial χ_j^2, and add the latter together to obtain the χ^2 statistic:

$$\chi_j^2 = (\text{actual}_j - m/8)^2 / (m/8)$$

$$\chi^2 = \sum_{j=1,8} \chi_j^2$$

The value of χ^2 so obtained is compared with tabulated values. Firstly, we find the number of degrees of freedom. If both mean and r.m.s. error were calculated from the data, there are 2 degrees of freedom less than the number of intervals. The tabulated values for 6 degrees of freedom are

Confidence level (%)	20	50	75	90	95	99	99.5	99.9
Tabulated χ^2		3.07	5.35	7.84	10.64	12.59	16.81	18.54	22.46

Any value of 12.6 or less shows that the distribution of residuals is to be considered normal at the 95% confidence level. The confidence level is to some extent arbitrary and not too important, since we really only want to identify residual distributions that are *not* normal. Alternatively, the number of residuals in each group may be plotted against magnitude as a histogram and compared visually with the corresponding normal distribution curve.

6.7.1.3 Parameter Values and Standard Deviations

A fit is only acceptable if the parameter values are themselves acceptable. This means that one should ideally have some idea as to what values to expect. These ideas will be in the form of bounds—the parameter value is expected to lie between a lower and an upper limit. A common lower limit is zero for any parameter that corresponds to a physical quantity. If the parameter value lies outside the expected bounds, suspicions should be roused. Maybe the estimated bounds are inaccurate; if not, the model is probably wrong. The standard deviations of the parameters are also very significant. A large standard deviation (e.g. more than 40% of the parameter value) implies that the parameter is not well defined by the experimental data, which could mean that the data are faulty or that the model is faulty. Either way, the model should not be accepted. If the quality of the data can be improved by further experiments, over a greater range, say, or with greater precision, it may be that the model can ultimately be accepted. The correlation coefficients are not as relevant to the question of model selection as whether the parameter values are meaningful. Parameters can also be tested against specific values, such as zero, or against each other, by the methods outlined in Sections 6.5 and 6.6.

6.7.1.4   Summary

There is no single criterion by which one may determine the acceptability of a model. All three aspects described above should be taken into account, and a model only accepted if the sum of squares is small, the residuals show no systematic trends and the parameter values and standard deviations are within reasonable bounds. Other aspects are discussed at the end of the following section.

6.7.2   CRITERIA FOR CHOOSING BETWEEN MODELS

It may happen that in trying to select a model, two or more candidates emerge that satisfy the criteria of acceptability. The choice between candidate models is rather difficult. If both models have the same number of parameters statistical tests will be useless, as the statistics depend on the distribution of experimental error, the number of data points and the number of parameters, and do not explicitly depend on the nature of the model.

When one model has more parameters than another it is natural to expect that the sum of squares, $S = \mathbf{r}^T \mathbf{W} \mathbf{r}$, will be smaller. For example, consider one fit with a polynomial of degree n and another with a polynomial of degree $n + 1$. The polynomial of higher degree will have additional flexibility, and will be able to follow the data points more precisely. In the limit when the degree of the polynomial is $m - 1$, an exact fit will be obtained.

We may apply an F test to the sums of squares of the two models. Let us call the S value for the model with the smaller number of parameters S_2 and the S value for the other model S_1. Presumably $S_2 > S_1$. Let the number of data points be m, the number of parameters in the smaller model be n and the number of extra parameters in the larger model be n_x. The ratio

$$\frac{\left(\dfrac{S_2 - S_1}{n_x}\right)}{\left(\dfrac{S_1}{m - n}\right)}$$

is distributed as F with n_x degrees of freedom in the numerator and $m - n$ degrees of freedom in the denominator. From the table of F we can find the level, α, at which the value of the ratio is significant. Clearing fractions, the ratio of variances can be tested:

$$\frac{S_2}{S_1} \approx \frac{n_x F(\alpha, n_x, m - n)}{m - n} + 1$$

For example, suppose we had fitted a set of 12 points with a straight line and a parabola, $m = 12$, $n = 2$, $n_x = 1$. The critical value of $F(0.005, 1, 10)$ is 12.826, so, if the variance ratio is less than 2.3 ($12.826/10 + 1$), we reject the hypothesis that the parabola gives a better fit at the 0.5% significance level. (Corresponding values are 2 and 1.5 at the 1 and 5% significance levels respectively.) The reader may be surprised that a near twofold reduction in the sum of squares is required for the parabola to give a significantly better fit. The ratio required is large in this instance because of the small number of degrees of freedom involved. With more data the number of degrees of freedom increases and the critical value of the variance ratio decreases to less than 2.

When the number of data points greatly exceeds the number of parameters we may use the approximation $m - n \approx \infty$, when (Section 6.2)

$$F(\alpha, r, m - n) \approx \chi^2(\alpha, r)/r$$

and test the variance ratio against the reduced χ^2.

An alternative to the variance ratio is provided by the generalized R factor, much used in crystallography. This is defined as

$$R = \left(\frac{(\mathbf{y}^{obs} - \mathbf{y}^{calc})^{\mathrm{T}} \mathbf{W} (\mathbf{y}^{obs} - \mathbf{y}^{calc})}{(\mathbf{y}^{obs})^{\mathrm{T}} \mathbf{W} \mathbf{y}^{obs}} \right)^{1/2}$$

It is not difficult to show that $(R_2^2 - R_1^2)/R_1^2 = (S_2 - S_1)/S_1$ and hence

$$\frac{R_2}{R_1} \approx \left(\frac{\chi^2(\alpha, n_x)}{m - n} + 1 \right)^{1/2}$$

A typical application of the variance ratio test is to see if a given model is improved by the addition of one or more extra parameters. There is overlap here with other forms of significance testing, such as testing that the additional parameters are significantly different from zero.

Having summarized some of the formal methods of hypothesis testing let us consider carefully the implications of using these methods. Statistical tests are predicated on the absence of systematic error. In practice one should view standard deviations as calculated by the least-squares method as *minimum* estimates of the true error. To understand this one has only to look in the literature at parameter values estimated by different methods (the *Stability Constants' Handbook* is a good source of such data). Reported values often differ from each other by margins that can be measured in terms of standard deviations. This can only result from the presence of systematic differences in the determinations—both errors in measurements and errors of approximation in theories. Knowing that systematic errors exist, it is unwise to try to extract too much information from a set of data: if one tries to squeeze too much juice from a lemon one obtains a lot of pith.

It should also be remembered that the discussion of statistics given in Section 6.4 is, strictly speaking, valid only for linear least-squares models. If the model is non-linear, the statistics are said to be *biased*, and all error estimates must be regarded as conservative.

Finally, it should never be forgotten that all the tests given here are based on the assumption that experimental errors are normally distributed. Tests based on other distributions can be developed, and indeed some may be found in specialist literature.

6.8 AN EXAMPLE: ANALYSIS OF ALUMINIUM AT TRACE LEVELS

In this section we look at a whole experiment which was designed to evaluate an analytical method for the determination of aluminium in water in the parts per billion (1 ppb = 10^{-9}) concentration range. The equipment used was a graphite furnace atomic absorption instrument. Samples were injected into the machine and the maximum meter reading, in arbitrary units, was recorded. In the first place the objective was to find out if the method could give a satisfactory linear calibration plot.

Each concentration was measured four times. The data are given in Table 6.3. Since the calibrant solutions were made up to predetermined concentrations this corresponds, statistically, to the Berkson case (Section 7.6.1). The four measurements were used to determine the mean response for that concentration and to estimate the error of the observation. Since only four measurements were taken this error estimate is likely to be pretty unreliable (Section 5.3), but it is better than having no error estimate at all. The data are to be fitted to the model

$$y = a + bx$$

Each observation is independent of the others, so there is no correlation and the weight matrix is diagonal. Each weight is taken to be the reciprocal of the error squared, so the data used for the calculation is given in Table 6.4. Three significant figures were used for the observations to be on the safe side, even though the experimental precision does not warrant them. Solution of

Table 6.3. Graphite furnace readings for aluminium determination

x/ppb	Raw data y^{obs}				Mean $\bar{y}$	Error s_y
15	132	102	110	97	110.25	15
20	133	137	114	120	126.0	11
25	144	113	211	154	153.25	42
30	225	254	217	154	212.5	42
35	212	160	249	135	189.0	52

Table 6.4. Data used to calculate the calibration line for aluminium determination

x	y^{obs}	w
15	110	0.004444
20	126	0.008264
25	153	0.000595
30	212	0.000567
35	189	0.000370

the normal equations

$$\begin{bmatrix} 0.0142 & 0.2768 \\ 0.2768 & 5.6408 \end{bmatrix} \begin{bmatrix} a \\ b \end{bmatrix} = \begin{bmatrix} 1.8113 \\ 36.4870 \end{bmatrix}$$

gives the results shown in Table 6.5, graphically shown in Figure 6.4.

Are the data a satisfactory set for a calibration plot? As far as the reduced sum of squares is concerned, the value is much lower than unity, which suggests that the model gives as good a representation of the data as can be expected. This is a reflection on the poor quality of the data. As far as the residuals are concerned, note particularly how the weighting scheme has affected the spread of residual values.

Is the intercept significantly greater than zero? We apply the t test to the ratio 31.8/21 and find that the hypothesis, $H_0: a > 0$, can be rejected at the 10% significance level but not at the 20% level. The implication of this test is that the 'blank' correction is not large, implying that the water used to make up the solutions is essentially free of dissolved aluminium.

Table 6.5. Least-squares parameters and residuals for the linear calibration in the aluminium determination, $y = a + bx$

$a = 31.8$, $s_a = 21$
$b = 4.91$, $s_b = 1.1$
$\rho_{ab} = 0.934$

x	y^{obs}	y^{calc}	$\sqrt{w}r$	wr^2
15	110	105.4	0.3049	0.0929
20	126	130.0	− 0.3601	0.1297
25	153	154.5	− 0.0365	0.0013
30	212	179.0	0.7850	0.6162
35	189	203.6	− 0.2801	0.0784
Sum				0.9167
$\mathbf{r}^T\mathbf{Wr}/m - n$				0.4583

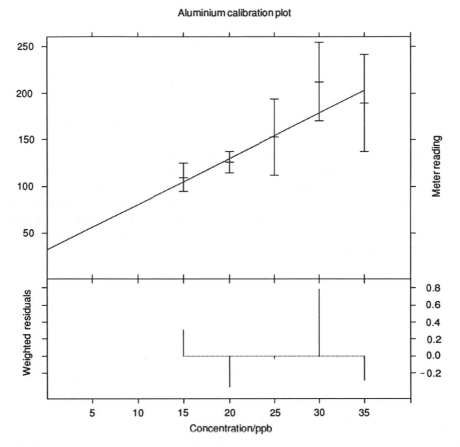

Figure 6.4. Calibration plot for aluminium determination. Upper box: experimental data and error bars of ±1 standard deviation, together with the least-squares straight line. Lower box: weighted residuals $r \times \sqrt{w}$

If this calibration plot were to be used for an aluminium determination, what would be the precision of the result? The calculation of concentration is simple:

$$x^{\text{calc}} = (y^{\text{obs}} - a)/b$$

We can use the error propagation formulae to get the error on x^{calc}:

$$\frac{s_x^2}{x^2} = \frac{s_y^2 + s_a^2}{(y - a)^2} + \frac{s_b^2}{b^2}$$

With reasonable assumptions (for example $x = 25$ ppb) this gives a relative error on the calculated concentration of about 30%. That may not seem very impressive until one recalls that a concentration of 25 ppb is equivalent to 25 mg per metric tonne!

How may the precision of the determination be improved? Clearly the precision of individual measurements needs to be improved. The average relative precision of the measurements quoted above is about 20%. It would need to be improved to somewhat less than 10% in order to yield a concentration precision of the order of 10%.

Postscript. The true objective of this experiment was to determine the level of aluminium in a soft drink marketed in an aluminium can. The soft drink is known to be markedly acidic, so it may be expected that some of the metal canister would be dissolved. Measurements on the soft drink gave very low readings, so it was concluded that the aluminium concentration was below the limit of detection. Looking at the formula for the error on the concentration we see that the error on a determination at 10 ppb would be near to 50%. Such a determination would have about a 50% chance of being equal to zero; the detection limit for this experiment is therefore about 10 ppb. One may conclude that the soft drink canister is well protected internally from attack by an acid solution.

7

Polynomials

7.1 INTRODUCTION

There is a vast literature on the fitting of polynomials to data, particularly in regard to straight line fitting. Much of this literature is concerned with ways of making hand calculations easier—in the days before electronic computers the word 'computer' meant a human being who did calculations! Nowadays many of the earlier techniques have been superseded with one notable exception: they are often used in hand calculating machines. For most of this chapter we will assume that the 'computer' is a programmable machine.

The general form of model that we shall adopt for the discussion of polynomial fitting is as follows:

$$y_i^{calc} = a_0 + a_1 x_i + a_2 x_i^2 + \cdots + a_{n-1} x_i^{n-1}$$

This represents a polynomial of degree $n-1$, which has n parameters $a_0, a_1, \ldots$, which may be designated the constant, linear, quadratic, etc., coefficients. It is often useful to perform a linear transformation of the independent variable:

$$z = T(x)$$

where T represents a linear transformation function. The transformation is performed before the calculation is commenced, and after the calculation is finished we perform a back-transformation:

$$x = T^{-1}(z)$$

together with an appropriate adjustment of the variance–covariance matrix of the parameters. In particular, if the data are equally spaced on the x axis by an increment h, the following linear transformation is very useful:

$$z = \frac{x - \bar{x}}{h}; \qquad x = hz + \bar{x}$$

where $\bar{x}$ is the average value of the x variable. When the data set contains m points the scaled variable, z, runs from $-(m-1)/2$ to $(m-1)/2$ in steps of 1, i.e. its values are

$$\frac{1-m}{2} \quad \frac{3-m}{2} \ldots 0 \ldots \frac{m-3}{2} \quad \frac{m-1}{2}$$

7.2 DIRECT SOLUTION

We consider first the case in which the errors on the observations are different, but uncorrelated. The variance-covariance matrix of the observations is diagonal:

$$\mathbf{M} = \mathrm{diag}(\sigma_i^2)$$

where σ_i^2 is the variance of the ith observation. The weight matrix is also diagonal, $\mathbf{W} = \mathbf{M}^{-1}$, and is easily factored: $\mathbf{w}^{\mathrm{T}}\mathbf{w} = \mathbf{W}$:

$$w_i = 1/\sigma_i$$

The ith row of the weighted Jacobian (Section 3.3) is

$$[J_i] = [w_i \quad w_i x_i \quad w_i x_i^2 \quad \cdots \quad w_i x_i^{n-1}]$$

The left-hand side of the normal equations can be built up by accumulating the pairwise products of the ith row such as w_i and $w_i x_i$, and the right-hand side by accumulating products such as w_i and $w_i y_i^{\mathrm{obs}}$. It follows that the normal equations are

$$\begin{bmatrix} \Sigma W_i & \Sigma W_i x_i & \Sigma W_i x_i^2 & \cdots & \Sigma W_i x_i^{n-1} \\ & \Sigma W_i x_i^2 & \Sigma W_i x_i^3 & \cdots & \Sigma W_i x_i^n \\ & & \Sigma W_i x_i^4 & \cdots & \cdots \\ \cdots & \cdots & \cdots & \cdots & \cdots \\ \text{symmetrical} & & & \cdots & \Sigma W_i x_i^{2n-2} \end{bmatrix} \begin{bmatrix} a_0 \\ a_1 \\ a_2 \\ \cdots \\ a_{n-1} \end{bmatrix} = \begin{bmatrix} \Sigma W_i y_i \\ \Sigma W_i x_i y_i \\ \Sigma W_i x_i^2 y_i \\ \cdots \\ \Sigma W_i x_i^{n-1} y_i \end{bmatrix}$$

$$\Sigma \equiv \sum_{i=1,m}$$

When the experimental errors are equal and uncorrelated there is a good case to be made for including the weights as above; this requires an estimate of the variance of the observations, but the effort required to obtain that estimate is amply repaid in terms of more tractable statistics (Chapter 6).

However, it is common practice to assign unit weights; this is justifiable when there is no hypothesis testing to be done, and results in somewhat simpler normal equations:

$$
\begin{bmatrix}
m & \Sigma x_i & \Sigma x_i^2 & \cdots & \Sigma x_i^{n-1} \\
 & \Sigma x_i^2 & \Sigma x_i^3 & \cdots & \Sigma x_i^n \\
 & & \Sigma x_i^4 & \cdots & \cdots \\
\cdots & \cdots & \cdots & \cdots & \cdots \\
\text{symmetrical} & & & \cdots & \Sigma x_i^{2n-2}
\end{bmatrix}
\begin{bmatrix}
a_0 \\ a_1 \\ a_2 \\ \cdots \\ a_{n-1}
\end{bmatrix}
=
\begin{bmatrix}
\Sigma y_i \\ \Sigma x_i y_i \\ \Sigma x_i^2 y_i \\ \cdots \\ \Sigma x_i^{n-1} y_i
\end{bmatrix}
$$

When there is correlation of the errors in the observations the normal equations cannot be accumulated from a single row of the Jacobian, but the whole Jacobian will be needed. It will be convenient to weight it $\hat{\mathbf{J}} = \mathbf{w}\mathbf{J}$ and to weight the observation $\hat{\mathbf{y}} = \mathbf{w}\mathbf{y}$.

The normal equations can be solved by standard methods of linear least squares (Section 3.5) in the general case, for the parameters and their errors. There are, however, a number of cases that can be solved more simply. These include the low-order polynomials ($n = 1, 2$) and polynomials up to degree five, as discussed under smoothing and differentiation below (Section 7.4).

The standard methods are satisfactory in the vast majority of cases. It is therefore worth while to use them even in the case of fitting straight lines, as only the one program is needed to process problems with any degree of polynomial. A little care is needed in formulating the problem. The x and y variables should be scaled so as to remove any large constant factors. This merely amounts to choosing the units so that the range of numbers is typically between -1000 and 1000. The reason for doing this is that products such as x^{2n-2} and $x^{n-1}y$ must be formed which might exceed the number capacity of the computer, being either too large (causing overflow) or too small (causing underflow). For example, if we were to seek to correlate the position of maximum absorption of visible light with some other variable, we would express the position is nanometres (visible light spans the range 400–700 nm) rather than the SI unit metres ($1 \text{ nm} = 10^{-9} \text{ m}$).

7.3

7.3.1   ORTHOGONAL POLYNOMIALS

The direct solution of the normal equations may not always be satisfactory. There is a problem inherent in the direct solution; for large n the normal equations become ill-conditioned with respect to inversion, i.e. the solution becomes extremely sensitive to small errors in the data and to round-off error in the calculation. Numerical analysts suggest that this problem becomes acute when $n \geqslant 7$. Therefore, orthogonal polynomials have been proposed as a way round the problem.

Before discussing this method we should ask the question, is the fitting of high-degree polynomials justified? The answer is—very rarely. The only situation in which a fit with a high-degree polynomial should be attempted is one in which there is a theoretical justification for the model. The author has been able to find only one instance in chemistry, the calculation of equilibrium constants from polarographic data. For empirical models there is now no need to use high-degree polynomials, as fitting with piecewise low-degree polynomials (spline functions; see Section 8.1) is much more satisfactory.

Perhaps the most useful property of the orthogonal polynomial method is that its parameters are independent of the degree of the polynomial. This means that one may test the addition of a term of degree n without any effect on the parameters obtained using polynomials up to degree $n-1$. For example, when trying out a linear and a quadratic fit the linear parameters are the same in both fits, and the quadratic parameter may be tested to see if it is significantly different from zero.

Orthogonal polynomials also have interesting theoretical properties which illuminate the whole process of polynomial fitting. The model for fitting with orthogonal polynomials is as follows:

$$y_i^{\text{calc}} = b_0 P_0(x_i) + b_1 P_1(x_i) + b_2 P_2(x_i) + \cdots + b_{n-1} P_{n-1}(x_i)$$

$P_j(x_i)$ is a polynomial of degree j in x_i. The normal equations for this model, assuming unit weights, are (omitting the x dependency)

$$
\begin{bmatrix}
\Sigma P_0^2 & \Sigma P_0 P_1 & \Sigma P_0 P_2 & \cdots & \Sigma P_0 P_{n-1} \\
 & \Sigma P_1^2 & \Sigma P_1 P_2 & \cdots & \Sigma P_1 P_{n-1} \\
 & & \Sigma P_2^2 & \cdots & \cdots \\
\cdots & \cdots & \cdots & \cdots & \cdots \\
\text{symmetrical} & & \cdots & \cdots & \Sigma P_{n-1}^2
\end{bmatrix}
\begin{bmatrix}
b_0 \\
b_1 \\
b_2 \\
\cdots \\
b_{n-1}
\end{bmatrix}
=
\begin{bmatrix}
\Sigma P_0 y \\
\Sigma P_1 y \\
\Sigma P_2 y \\
\cdots \\
\Sigma P_{n-1} y
\end{bmatrix}
$$

The orthogonality of the polynomials is defined by setting all the off-diagonal terms to zero:[*]

$$\Sigma P_j P_k = \sum_{i=1,m} P_j(x_i) P_k(x_i) = 0$$

Notice that orthogonality is defined in terms of the values of the independent variable; if a data point is added or removed the orthogonal polynomials

---

[*] With unequal but uncorrelated errors the first normal equation would be

$$[\Sigma W_i P_0(x_i)^2] b_0 + [\Sigma W_i P_0(x_i) P_1(x_i)] b_1 + \cdots = \Sigma W_i P_0(x_i) y_i$$

and the orthogonality conditions become

$$\Sigma W_i P_j(x_i) P_k(x_i) = 0$$

will change. With this definition, the solutions to the normal equations can be written down directly, thus eliminating any possibility of ill-conditioning:

$$b_j = \frac{\Sigma P_j(x_i) y_i^{obs}}{\Sigma P_j(x_i)^2}$$

An example of the explicit solution of these normal equations is given in Section 7.5.4.

The properties of orthogonal polynomials have been discussed in detail by Guest (1961), and only the relevant results will be quoted here. The orthogonal polynomials can be expressed as a series in powers of x, and vice versa:

$$P_j(x) = \sum_{k=0,j} \beta_{kj} x^k$$

$$x^j = \sum_{k=0,j} \alpha_{kj} P_k(x)$$

Guest shows how the coefficients α and β are related to the L and U factors in the LU factorization of the normal equations (called by Guest the Gauss–Doolittle scheme). From these relationships he shows how the parameters b are related to the parameters a of the direct solution:

$$a_j = \sum_{k=j,n-1,} \beta_{jk} b_k$$

Fitting by orthogonal polynomials is completely equivalent to fitting by the direct method. Indeed, once the coefficients are established, the polynomial may be written out explicitly in terms of the a parameters of the direct method.

If the coefficients had to be obtained by factorization of the normal equations there would be no advantage in using this method. Instead, the orthogonal polynomials may be calculated by means of recurrence relationships which allow the polynomial value to be calculated at any point x_h:

$$P_{j+1}(x_h) = x_h P_j(x_h) - \frac{\Sigma x_i P_j^2(x_i)}{\Sigma P_j(x_i)^2} P_j(x_h) - \frac{\Sigma P_j(x_i)^2}{\Sigma P_{j-1}^2(x_i)} P_{j-1}(x_h)$$

The summations in this formula extend over all the data points. It is also recommended that the x variable should be scaled to the range $-2 < z < 2$ so that successive polynomials have similar orders of magnitude. The recurrence relations use the definitions

$$P_0 = 1; \qquad P_1 = x - \bar{x}: \qquad \bar{x} = \Sigma x_i / m$$

7.3.2   EQUALLY SPACED DATA POINTS

When the data points are equally spaced by an increment h it is convenient to employ the change of variable (Section 7.1):

$$z = \frac{x - \bar{x}}{h}$$

With this change of variable the summations involved in the recurrence relations can be easily evaluated in closed form and hence the coefficients of the
orthogonal polynomials can also be evaluated in closed form:

$$\beta_{kk} = 1$$
$$\beta_{kj} = 0 \text{ for } k + j \text{ odd}$$
$$\beta_{k+1,j+1} = \beta_{kj} - \frac{j^2(m^2 - j^2)}{4(4j^2 - 1)} \beta_{k+1,j-1}$$

The coefficients for polynomials up to degree 9 have been tabulated by Guest
(1961); those for polynomials up to degree 6 are given in Table 7.1.
To illustrate the use of this table let us consider a simple example: fitting a
quadratic orthogonal polynomial. The model is

$$y_i = b_0 P_0(z_i) + b_1 P_1(z_i) + b_2 P_2(z_i)$$

Suppose that there are five equally spaced points, $z = -2, -1, 0, 1, 2$ ($m = 5$).
The three orthogonal polynomials are obtained by multiplying the coefficients
in the first three rows of the table by the appropriate power of z and adding
the products together:

$$P_0 = 1$$
$$P_1 = z - \bar{z} = z \qquad (\bar{z} = 0)$$
$$P_j(z) = \sum_{k=0,j} \beta_{kj} z^k$$
$$P_2 = -(m^2 - 1)/12 + 0z + 1z^2 = z^2 - 2$$

The ith row of the Jacobian is

$$[J_i] = [1 \quad z_i \quad z_i^2 \quad -2]$$

The normal equations are obtained from this, bearing in mind that the
polynomials are orthogonal so that the normal equations matrix is diagonal:

$$\begin{bmatrix} 5 & 0 & 0 \\ 0 & 10 & 0 \\ 0 & 0 & 14 \end{bmatrix} \begin{bmatrix} b_0 \\ b_1 \\ b_2 \end{bmatrix} = \begin{bmatrix} 1 & 1 & 1 & 1 & 1 \\ -2 & -1 & 0 & 1 & 2 \\ 2 & -1 & -2 & -1 & 2 \end{bmatrix} \begin{bmatrix} y_{-2} \\ y_{-1} \\ y_0 \\ y_1 \\ y_2 \end{bmatrix}$$

The solution to these equations is

$$b_0 = (\quad y_{-2} + y_{-1} + \quad y_0 + y_1 + \quad y_2)/5$$
$$b_1 = (-2y_{-2} - y_{-1} \qquad + y_1 + 2y_2)/10$$
$$b_2 = (\quad 2y_{-2} - y_{-1} - 2y_0 - y_1 + 2y_2)/14$$

Table 7.1 Coefficients β_{kj}: $P_j(z) = \Sigma \beta_{kj} z^k$ for m equally spaced observations

j	0	1	2	3	4	5	6
0	1						
1	0	1					
2	$-\dfrac{(m^2-1)}{12}$		1				
3		$-\dfrac{(3m^2-7)}{20}$		1			
4	$\dfrac{3(m^2-1)(m^2-9)}{560}$		$-\dfrac{(3m^2-13)}{14}$		1		
5		$\dfrac{15m^4-230m^2+407}{1008}$		$-\dfrac{5(m^2-7)}{18}$		1	
6	$-\dfrac{5(m^2-1)(m^2-9)(m^2-25)}{14784}$		$\dfrac{5m^4-110m^2+329}{176}$		$-\dfrac{5(3m^2-31)}{44}$		1
k	0	1	2	3	4	5	6

At this stage we may calculate the *a* parameters:

$$a_0 = b_0 - (m^2 - 1)/12b_2$$
$$= (-3y_{-2} + 12y_{-1} + 17y_0 + 12y_1 - 3y_2)/35$$
$$a_1 = b_1$$
$$a_2 = b_2$$

The errors on the *b* parameters are very easy to calculate as the normal equations matrix is diagonal and therefore simple to invert:

$$\sigma^2(b_0) = \frac{\mathbf{r}^T\mathbf{r}}{5(m-3)}; \qquad \sigma^2(b_1) = \frac{\mathbf{r}^T\mathbf{r}}{10(m-3)}; \qquad \sigma^2(b_2) = \frac{\mathbf{r}^T\mathbf{r}}{14(m-3)}$$

The error on a_0 will be obtained by error propagation:

$$\sigma^2(a_0) = \sigma^2(b_0) + (m^2 - 1)^2/144\sigma^2(b_2)$$

Alternatively, the error on a_0 can be found from the coefficients of *y* to be $17/35\sigma^2(y)$, i.e. about half the error on an observation. This approach to errors is taken up again in Section 7.4.3.

To facilitate the use of orthogonal polynomials on equally spaced points it is also useful to have a closed formula for the sum of the squared values, as these are the entries in the normal equations matrix. Guest gives the following formula:

$$\Sigma P_j(z)^2 = \frac{j!^4}{(2j)!(2j+1)!} \, m(m^2 - 1)\cdots(m^2 - j^2)$$

where *j*! stands for factorial *j*. Evidently this number can become very large as *m* increases for polynomials of high degree. For example, when fitting a sextic to 100 data points

$$P_6(z) = 3.370\,206\,27 \times 10^8 + 2.834\,81 \times 10^5 z^2 - 3.4056 \times 10^3 z^4 + z^6$$

$$\Sigma P_6(z)^2 = \frac{6!^4}{12! \times 13!} \, 10^2(10^4 - 1)\cdots(10^4 - 36)$$

$$\approx 8.92 \times 10^{18}$$

These numbers show that although the use of orthogonal polynomials eliminates the problem of ill-conditioned normal equations, there still remains some difficulty associated with round-off errors—the errors introduced when adding or subtracting two numbers that are of very different orders of magnitude. The only safe procedure is to employ multilength arithmetic, i.e. carry extra significant figures in the calculation.

7.4

7.4.1 SMOOTHING, DIFFERENTIATION AND INTEGRATION BY CONVOLUTION

It was noted in Section 7.3.2 that the expressions for the orthogonal polynomial become much simpler when the data are equally spaced. It is also true that the solution of the direct problem is hugely simplified when the data are equally spaced. Equally spaced data are common nowadays with microprocessor-controlled instruments that produce digital output; commonly data are output at constant time increments. Therefore the special formulae that can be derived for the equally spaced data may find wide application.

Let the spacing between data points be h. With the change of variable

$$z = \frac{x - \bar{x}}{h}; \qquad \bar{x} = \Sigma x / m$$

all the sums of odd powers of z become equal to zero, and the normal equations become

$$
\begin{bmatrix}
m & 0 & \Sigma z^2 & 0 & \cdots \\
0 & \Sigma z^2 & 0 & \Sigma z^4 & \cdots \\
\Sigma z^2 & 0 & \Sigma z^4 & 0 & \cdots \\
0 & \Sigma z^4 & 0 & \Sigma z^6 & \cdots \\
\cdots & \cdots & \cdots & \cdots & \cdots
\end{bmatrix}
\begin{bmatrix}
a_0 \\
a_1 \\
a_2 \\
a_3 \\
\cdots
\end{bmatrix}
=
\begin{bmatrix}
\Sigma y \\
\Sigma zy \\
\Sigma z^2 y \\
\Sigma z^3 y \\
\cdots
\end{bmatrix}
$$

These equations factor into two separate sets of equations:

$$
\begin{bmatrix}
m & \Sigma z^2 & \cdots \\
\Sigma z^2 & \Sigma z^4 & \cdots \\
\cdots & \cdots & \cdots
\end{bmatrix}
\begin{bmatrix}
a_0 \\
a_2 \\
\cdots
\end{bmatrix}
=
\begin{bmatrix}
\Sigma y \\
\Sigma z^2 y \\
\cdots
\end{bmatrix}
$$

$$
\begin{bmatrix}
\Sigma z^2 & \Sigma z^4 & \cdots \\
\Sigma z^4 & \Sigma z^6 & \cdots \\
\cdots & \cdots & \cdots
\end{bmatrix}
\begin{bmatrix}
a_1 \\
a_3 \\
\cdots
\end{bmatrix}
=
\begin{bmatrix}
\Sigma zy \\
\Sigma z^3 y \\
\cdots
\end{bmatrix}
$$

The parameters can be obtained from the analytical formula

$$\mathbf{a} = [(\mathbf{J}^T\mathbf{J})^{-1}\mathbf{J}^T]\mathbf{y} = \mathbf{C}\mathbf{y}$$

The rows of the matrix $\mathbf{C}$ constitute coefficients by which the observations are to be multiplied and summed:

$$a_{ki} = \sum_{j=-p,p} C_{kj} y_{i+j}; \qquad p = (m-1)/2$$

a_{ki} represents the value of the kth parameter at the ith data point. If the number of points fitted, m, is less than the total number, the process may be applied to the first m data points, then to m points starting at the second, and so forth. The process as a whole is called a *convolution* of the data because of the form of the equation above. Convolution is discussed again in Section 9.3.

There is one quirk in the system that arises out of the data points being equally spaced. For even-numbered derivatives (smoothing is counted as the zeroth derivative) the same results are obtained from polynomials of degree 0 or 1, 2 or 3, etc., and for the odd-numbered derivatives the same results are obtained from polynomials of degree 1 or 2, 3 or 4, etc.

The convolution procedure provides a method for smoothing, differentiating and integrating the data numerically. From the form of the polynomial

$$y = a_0 + a_1 z + a_2 z^2 + a_3 z^3 + \cdots = a_0 \text{ at } z = 0$$

$$\frac{dy}{dz} = a_1 + 2a_2 z + 3a_3 z^2 + \cdots \qquad = a_1 \text{ at } z = 0$$

$$\frac{d^2 y}{dz^2} = 2a_2 + 6a_3 z + \cdots \qquad = 2a_2 \text{ at } z = 0$$

Therefore, the least-squares parameters themselves furnish estimates of the derivatives at the centre of the data subset fitted by the polynomial. If the number of points in the subset is an odd number, the parameters provide estimates of the derivatives at the central point (the $(n-1)$th derivative is estimated for the whole of the subset). If m is an even number the derivatives are estimated at the mid-point between the two central points.

To make the procedure clearer let us work it out for the case of fitting a quadratic polynomial to a set of five data points ($z = -2, -1, 0, 1, 2$). The normal equations matrix and its inverse are

$$\mathbf{J}^{\mathrm{T}}\mathbf{J} = \begin{bmatrix} 5 & 0 & 10 \\ 0 & 10 & 0 \\ 10 & 0 & 34 \end{bmatrix}, \quad (\mathbf{J}^{\mathrm{T}}\mathbf{J})^{-1} = \begin{bmatrix} 34/70 & 0 & -10/70 \\ 0 & 1/10 & 0 \\ -10/70 & 0 & 5/70 \end{bmatrix}$$

Calculation of the inverse is made easy by the way the normal equations matrix can be factored. To form the coefficient matrix $\mathbf{C}$ we left-multiply the transposed Jacobian by $(\mathbf{J}^{\mathrm{T}}\mathbf{J})^{-1}$:

$$\mathbf{C} = \begin{bmatrix} 34/70 & 0 & -10/70 \\ 0 & 1/10 & 0 \\ -10/70 & 0 & 5/70 \end{bmatrix} \begin{bmatrix} 1 & 1 & 1 & 1 & 1 \\ -2 & -1 & 0 & 1 & 2 \\ 4 & 2 & 0 & 2 & 4 \end{bmatrix}$$

$$= \begin{bmatrix} -6/70 & 24/70 & 34/70 & 24/70 & -6/70 \\ -2/10 & -1/10 & 0 & 1/10 & 2/10 \\ 10/70 & -5/70 & -10/70 & -5/70 & 10/70 \end{bmatrix}$$

$$a_{0i} = (-3y_{i-2} + 12y_{i-1} - 17y_i + 12y_{i+1} - 3y_{i+2})/35$$

$$a_{1i} = (-2y_{i-2} \quad - y_{i-1} \qquad\qquad + y_{i+1} + 2y_{i+2})/10$$

$$a_{2i} = \quad (2y_{i-2} \quad - y_{i-1} \ -2y_i \quad - y_{i+1} + 2y_{i+2})/14$$

The results are, of course, the same as were found when using orthogonal polynomials (Section 7.3.2).

To obtain the derivatives as a function of the original independent variable x we must apply a back-transformation:

$$x = \bar{x} + zh$$

$$\frac{dy}{dx} = \frac{dz}{dx}\frac{dy}{dz} = \frac{1}{h}\frac{dy}{dz}$$

$$\frac{d^k y}{dx^k} = \frac{1}{h^k}\frac{d^k y}{dz^k}$$

At this stage let us try to obtain an overview of the convolution process. The appearance is deceptive as the convolution formula appears to have nothing to do with least-squares data-fitting. However, the fundamental process is one of fitting a polynomial of degree $n - 1$ to a set of m data points. This is true whether one is using convolution to obtain derivatives or to smooth the data. The calculated data always refers to the parameters of the best-fit polynomial over the subset of m data points. For this reason it is recommended that the smoothing function should always be calculated, and the difference between observed data and smoothed data should be displayed. An example is shown in Figure 7.1. The display is divided into three areas. The upper box shows the residuals, $y^{obs} - y^{smoothed}$, the middle box the original data and the lower box shows the derivative. This display makes it clear that the derivative shown relates to the *smoothed* curve, not the original data.

One minor drawback exists when the convolution method is used to calculate the derivatives at the central point. This is that the first and last $(m - 1)/2$ points are not calculated and are therefore lost. This can easily be overcome by calculating the points using the complete polynomial expression, a procedure that requires all the a coefficients to be calculated for the first and last m data points. An alternative procedure is to redefine the subsidiary variable:

$$z = \frac{x - x_1}{h}$$

With this scaling $z = 0, 1, \ldots, m - 1$. A new set of convolution coefficients may be calculated to give derivatives at the first point. For example, for a five-point quadratic convolution they are

$$a_{0i} = (\quad 62y_i + 18y_{i+1} - \ 6y_{i+2} - 10y_{i+3} + \ 6y_{i+4})/70$$

$$a_{1i} = (-54y_i + 13y_{i+1} + 40y_{i+2} + 27y_{i+3} - 26y_{i+4})/70$$

$$a_{2i} = (\quad 10y_i - \ 5y_{i+1} - 10y_{i+2} - \ 5y_{i+3} + 10y_{i+4})/70$$

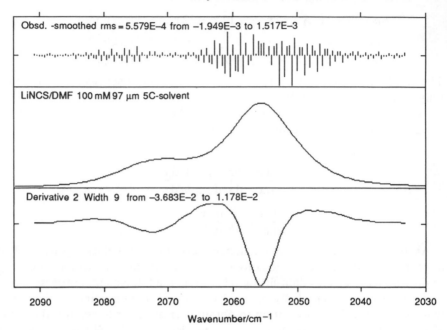

Figure 7.1. Smoothing and differentiation by convolution. Upper box: residuals obtained by smoothing. Middle box: experimental data. Lower box: second derivative of the smoothing polynomials

Note that with this scaling for z the normal equations do not factorize into two sets of equations, so the coefficients given above were obtained by solving the three normal equations simultaneously. The convolution coefficients for last-point smoothing are the same as those for first-point smoothing; to obtain estimates of the derivatives at the last point in a set one applies these coefficients to the data in reverse order.

7.4.2 ALGEBRAIC EXPRESSIONS FOR THE CONVOLUTION COEFFICIENTS

For polynomials of degree 3 or less the calculation of the convolution coefficients involves the solutions of only two simultaneous equations. It is therefore straightforward, though tedious, to derive analytical expressions for the convolution coefficients. The following identities are useful:

$$\Sigma z^2 = m(m^2 - 1)/12$$
$$\Sigma z^4 = m(m^2 - 1)(3m^2 - 7)/240$$
$$\Sigma z^6 = m(m^2 - 1)(3m^4 - 18m^2 + 31)/1344$$

The summations are between $z = -(m - 1)/2$ and $z = (m - 1)/2$. The 2×2 matrices can be inverted simply:

$$\begin{bmatrix} a & b \\ b & c \end{bmatrix}^{-1} = \begin{bmatrix} c/d & -b/d \\ -b/d & a/d \end{bmatrix}, \qquad d = ac - b^2$$

One way of generating the convolution coefficients is to calculate the sums, the inverse matrix as above, and multiply this on the right by the transposed Jacobian, whose ith row is

$$[J_i] = [1 \quad z_i \quad z_i^2 \quad \ldots]; \qquad z_i = i - (m + 1)/2$$

Thus, the coefficients for smoothing by a quadratic/cubic polynomial are

$$C_{0j} = \frac{-\Sigma z^4 + j^2 \Sigma z^2}{m \Sigma x^4 - (\Sigma z^2)^2}; \qquad j = \frac{1 - m}{2} \ldots \frac{m - 1}{2}$$

However, since the summations can be written down as expressions one might alternatively wish to use the fully derived formulae. The following apply to the use of a cubic polynomial:

$$C_{0j} = \frac{(3m^2 - 7 - 20j^2)/4}{m(m^2 - 4)/3}$$

$$C_{1j} = \frac{5(3m^4 - 18m^2 + 31)j - 28(3m^2 - 7)j^3}{m(m^2 - 1)(3m^4 - 39m^2 + 108)/15}$$

$$C_{2j} = \frac{12mj^2 - m(m^2 - 1)}{m^2(m^2 - 1)(m^2 - 4)/15}$$

$$C_{3j} = \frac{(3m^2 - 7)j + 20j^3}{m(m^2 - 1)(3m^4 - 39m^2 + 108)}$$

As discussed below, the cubic polynomial is probably the most useful for general purposes, so these coefficients are likely to be used the most often.

The algebra for higher-degree polynomials becomes increasingly tedious; the simplest way to get expressions is via orthogonal polynomials. Fortunately all the relevant expressions have been published by Madden (1978); note that the m in Madden's expressions is $(m - 1)/2$ in terms of the usage of this book.

Note. Although the convolution coefficients can be expressed as a quotient of integers this is not a convenient form for use with computers. The reason is that the integers involved in the calculation become very large, and will usually overflow the integer capacity of the computer for modest values of m (e.g. $m = 21$ when calculating the coefficients for the first derivative for a cubic polynomial on a machine with four-byte integer precision). It is more convenient to use floating-point arithmetic throughout, in which case the convolution coefficients will all be less than one.

7.4.3 SMOOTHING—CHOOSING THE BEST CONVOLUTION FUNCTION

We may examine the smoothing process to try to find the optimal choice of polynomial degree and the number of points that should be used. Firstly, let us assume that the piecewise polynomial approximation is a good one, i.e. that the errors in the polynomial fit are negligible in comparison to the experimental errors on the data. In this case, smoothing will merely remove noise, i.e. experimental error.[*] To simplify the calculation let us assume that the variance at each data point is the same, σ^2, and that there is no correlation between experimental errors. By the law of propagation of errors (Section 2.2), the error on the ith smoothed data point is given by

$$\sigma^2(a_{0i}) = \sum_{j=-p,p} C_{0j}^2 \sigma^2(y_{i+j}) = \sigma^2 \Sigma C_{0j}^2$$

It can be shown (Gans and Gill, 1983) that the sum of the squares of the convolution coefficients for mid-point smoothing is equal to the value of the central convolution coefficient, C_{00}. From this fact, and using the analytical form of the normal equations, we can calculate the variance of the smoothed data point, as follows:

Degree 0, 1:
$$\sigma^2(a_{0i}) = \frac{1}{m}\sigma^2$$

Degree 2, 3:
$$\sigma^2(a_{0i}) = \frac{3(3m^2 - 7)}{4m(m^2 - 4)}\sigma^2 \approx \frac{9}{4m}\sigma^2$$

Degree 4, 5:
$$\sigma^2(a_{0i}) = \frac{15(15m^4 - 230m^2 + 407)}{64m(m^2 - 4)(m^2 - 16)} \approx \frac{225}{64m}\sigma^2$$

The standard deviation, expressed as a percentage of the standard deviation of the original data point, is plotted in Figure 7.2. Curve (a) is for polynomials of degree 0 and 1, curve (b) is for the quadratic/cubic case and curve (c) is for degrees 4 and 5. This shows that the lower-order polynomial removes noise more efficiently and that the amount remaining decreases quite rapidly at first and then more slowly. Therefore, from the point of view of noise removal, one should choose the lowest-degree polynomial and the largest number of points in the convolution. The order of magnitude of the noise reduction should also be noted. For example, use of a 25-point smoothing function leads to noise

---

[*] For this reason the convolution process may be described as a filtering process and the convolution function as a filter. The smoothing function filters noise out of the data.

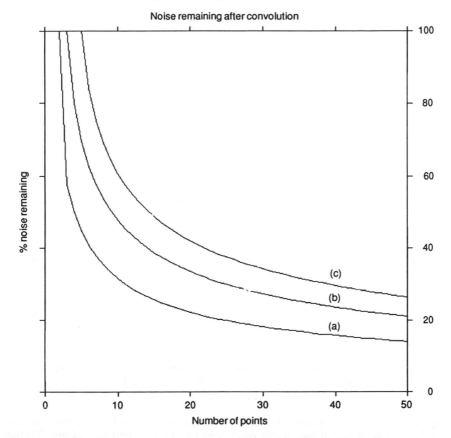

Figure 7.2. Noise reduction for smoothing by convolution: (a) polynomials of degree 0 or 1, (b) polynomials of degree 2 or 3, (c) polynomials of degree 4 or 5. The curves show the percentage remaining after convolution of the standard deviation of the noise

levels of approximately 20, 30 and 37.5% of the original with polynomials of degrees 1, 3 and 5 respectively.

The nature of the noise removed can be examined by means of Fourier transform techniques. The convolution functions that we have been discussing behave as low-pass filters, so that the noise removed is predominantly high-frequency noise. For more details see Section 9.5.

Noise removal is not, however, the only consideration. Some distortion is inevitable in all cases where the ideal (noise-free) data cannot be represented by a polynomial. To illustrate this the residuals obtained by smoothing a noise-free Lorentzian with a quadratic/cubic function are shown in Figure 7.3.

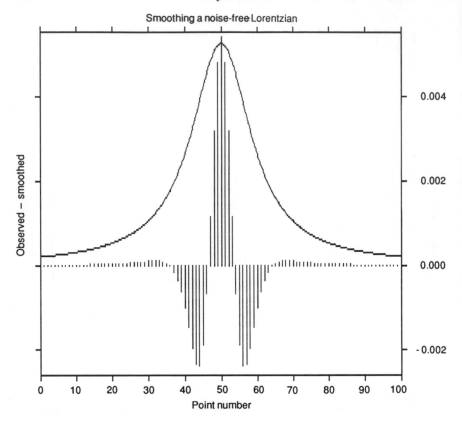

Figure 7.3. Smoothing a noise-free Lorentzian using quadratic/cubic polynomials. Solid line: Lorentzian of half-width equivalent to 20 points and unit height. Vertical bars: residuals obtained after convolution with an 11-point quadratic/cubic smoothing function. The scale for residuals is at the right

Distortion is evident over most of the Lorentzian band, with the maximum distortion in this case being at the position of the band maximum.* Distortion is further discussed in Appendix 7.

The amount of distortion is reduced by increasing the degree of the polynomial and/or by decreasing the convolution width. In other words, the requirements for minimizing distortion are the opposite of what is needed to remove noise from the data. In practice, therefore, a compromise is needed. Let us consider each type of polynomial in turn.

---

* With symmetric functions such as the Lorentzian it is convenient to define the *smoothing ratio* as the ratio of the half-width of the band to the width of the convolution function. In Figure 7.3 the band half-width was equivalent to 20 points and an 11-point smoothing function was used, corresponding to a smoothing ratio of 20/11, or 1.82.

The degree 1 case is the method of the moving average. The convolution coefficients are all $1/m$. Since the function fitted is a straight line, it will not be very useful for data that has strong curvature. It is unlikely, then, to be of much service in general. If the data interval is small compared to the rate of change of the dependent variable it may be worth while to use a three-point moving average in the first instance, as this will reduce the noise to about 60% of its original value.

Fitting with a cubic is the same as fitting with a quadratic for the purpose of smoothing, so let us consider this case next. The cubic polynomial may have an inflection point, and so is suitable for fitting data made up of functions that have inflection points, such as Gaussians, Lorentzians, trigonometric functions, etc. The largest distortion will occur wherever the polynomial approximation is poorest, which with Gaussians and Lorentzians is at the peak maxima.

Fitting with a quintic may give less distortion, but the noise removal is not so good. The quintic function may have two maxima and two minima. It is therefore unlikely to be required.

For general purposes it appears that fitting with a cubic is likely to be satisfactory. Let us now consider the choice of the number of points to include in the convolution. As we have seen above, as the number of points is increased the amount of noise removed also increases. This is illustrated in Figure 7.4(a). The noise removed is simply 1—the noise remaining. The amount of 'noise' generated by smoothing a Lorentzian is shown in Figure 7.4(b). This increases as the number of convolution points increases. In general, any function can be expanded about a given point as a polynomial, but the range over which a low-order approximation is good depends on the nature of the function, and most functions will eventually fail to be well approximated.

If we plot the sum of noise removed and distortion added as a function of the number of convolution points, as in Figure 7.4(c), the curve has an inflection point where the acceleration of the amount of noise removed is equal to the acceleration of 'noise' added by distortion. This inflection point gives an optimal width for the convolution function; a wider convolution function would give better noise removal at the expense of more distortion. An empirical plot of the sum of $(y_i^{obs} - a_{0i})^2$ will have the same shape as in Figure 7.4(c), so the optimal width for convolution may be chosen empirically.

The position of the inflection point is dependent on the relative magnitudes of experimental error and smoothing distortion. The idea behind choosing the inflection point is that distortion should not be greater than experimental error.

The residuals $(y_i^{obs} - a_{0i})$ should always be plotted to see if distortion is significant. If it is, there will be a clump of residuals of the same sign just as in Figure 7.3 superimposed on the residuals which represent noise removed. If distortion is to be avoided the width of the convolution function must be reduced until the residuals show no systematic trends.

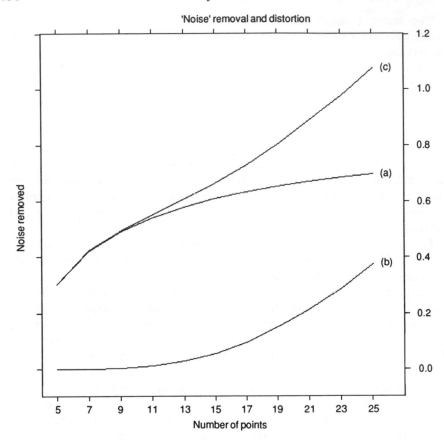

Figure 7.4. Noise removed and generated by smoothing a Lorentzian band: (a) noise removed by smoothing, (b) 'noise' generated by distortion, (c) sum of (a) and (b)

For some applications the amount of distortion is not so important, so long as it is reproducible. There are instances in analytical chemistry where an analyte is measured against a calibration standard. Smoothing can be used as long as it is applied consistently to both analyte and calibrant; i.e. the same degree of polynomial and same number of convolution points are used for both (see Enke and Niemann, 1976).

There are not many instances when smoothing is used *per se*. Most scientists would probably say that smoothing by itself is merely cosmetic. Indeed, some would argue that it is dangerous because it masks the true nature and magnitude of experimental error. We have discussed smoothing here because it is an essential concomitant to the method of numerical differentiation, which does have widespread application.

One possible application is in data compression. Suppose that a large number of data points has been collected and that the data changes only slowly with the independent variable. It may be worth while to reduce the number of data points by a factor of 2, whilst also improving the precision of the compressed data. Thus to use three-point smoothing, the formula to adopt would be

$$y_{i/2}^{\text{compressed}} = \tfrac{1}{3}(y_{i-1} + y_i + y_{i+1})^{\text{obs}}; \qquad i = 2, 4, 6, \ldots, m$$

The precision of the compressed data would be improved by a factor of about 1.7. This may be thought of as a form of coaddition (Section 5.4.3) which uses one data set rather than three, with consequent saving in time. However, the two processes are not equivalent since coaddition does not imply that the data can be approximated by a straight line as does the formula above. Given that smoothing inevitably introduces distortion, this form of data compression should only be used when distortion is immaterial.

Two final comments need to be made. There are a number of software packages that include the convolution method for smoothing and differentiation, but which offer a limited number of choices in regard to polynomial degree and convolution width, often in a disguised form such as 'damping'. Obviously care must be exercised when using such a package! Lastly, the presentation given here does not follow the historical development of the subject. The convolution method was popularized by Savitzsky and Golay (1964). They published tables of the convolution coefficients which were widely used for a while. However, the method goes back much further in time, probably to the nineteenth century. Guest (1961) gives a small table of smoothing coefficients, and also expounds a general method for calculating the smoothing coefficients using the corresponding orthogonal polynomials. It is only right that this section should be concluded by acknowledging these historical sources.

7.4.4 NUMERICAL DIFFERENTIATION

We shall now consider ways of obtaining an estimate of the derivatives of an experimental curve. The data consist of the measured values y_i at equal increments, h, along the x axis. It is important to recognize that the derivatives are inherently more noisy than the original data. To see how this comes about let us take a simple case, fitting a cubic to five data points:

$$\frac{dy_i}{dx} \approx \frac{y_{i+2} - 8y_{i+1} + 8y_{i-1} - y_{i-2}}{12h}$$

The *value* of the first derivative is much smaller than the measured values because it is the sum of differences between similar numbers, but, assuming the data have equal variance σ^2, the *error* on the derivative is $\sqrt{(130/144)}\, h^2\sigma$. It follows that the ratio of derivative error to value is much larger than the ratio of observed error to value. Therefore it is essential to use smoothing when calculating the derivative.

The variance on the derivative is $\sigma^2 \Sigma_j C_{1j}^2$. Evaluation of the relevant expressions gives the variance on the first derivative as

$$\text{Degree } 1, 2: \qquad \sigma^2(a_{1i}) = \frac{12}{m(m^2 - 1)} \, \sigma^2 \approx \frac{12}{m^3} \, \sigma^2$$

$$\text{Degree } 3, 4: \qquad \sigma^2(a_{1i}) = \frac{(3m^4 - 18m^2 + 31)/112}{m(m^2 - 1)(3m^4 - 39m^2 + 108)/8400} \approx \frac{75}{m^3} \, \sigma^2$$

$$\text{Degree } 5, 6: \qquad \sigma^2(a_{1i}) \approx \frac{260}{m^3} \, \sigma^2$$

This variance decreases approximately in inverse proportion to the cube of the number of data points, in contrast to the case for smoothing, which decreases in inverse proportion to m. It needs to do so considering the fact that the ratio of derivative error to value is much larger. For the second derivative it may be shown that the variance decreases approximately in inverse proportion to the fifth power of the number of data points. The convolution method is therefore very useful for numerical differentiation of experimental curves.

We have seen that the cubic polynomial probably offers the best compromise between smoothing and distortion, so the derivatives should be calculated in most cases with a convolution function based on the cubic polynomial. A formula for the convolution coefficients has been given in Section 7.4.2.

The third derivative is a constant when the polynomial is cubic, so third and higher derivatives would seem to require the use of a polynomial of degree greater than 3. This is not inevitable, as there is another option available for the higher derivatives. This involves repeated use of a first derivative function. For example, the second derivative may be calculated directly from the polynomial, or by differentiating an estimate of the first derivative. In that way all smoothing and differentiation may be performed with cubic polynomials. Repeated convolutions are discussed in more detail in Section 7.4.5.

At this point it is appropriate to outline some of the applications of numerical differentiation. Some are based on the notion that the experimental curve is made up of a sum of signals that behave differently under differentiation in such a way that the derivative curve 'resolves' the components contributing to the data. For example, a common task in analytical chemistry is to measure the height of a band superimposed on a curved background, as shown in Figure 7.5. This set of 310 data points was constructed with a quadratic baseline added to a Lorentzian band. The second derivative shows a band on an

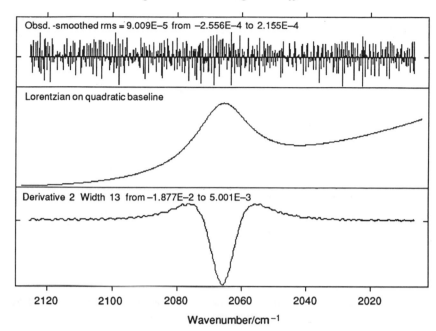

Obsd. -smoothed rms = 9.009E–5 from –2.556E–4 to 2.155E–4

Lorentzian on quadratic baseline

Derivative 2 Width 13 from –1.877E–2 to 5.001E–3

| 2120 | 2100 | 2080 | 2060 | 2040 | 2020 |

Wavenumber/cm^{-1}

Figure 7.5. Removal of the effect of a curved background. Upper box: residuals obtained after smoothing. Middle box: a simulated Lorentzian on a quadratic baseline. Lower box: second derivative of the smoothing polynomials

almost flat baseline.[*] Since the maximum depth of the second derivative of a Lorentzian band is proportional to the maximum height of the band, effectively the height of the band can be measured from the derivative without interference from the curved baseline. The same would be true if the band were Gaussian, so numerical differentiation provides a powerful tool for the analysis of this kind of data.

Another application of resolution enhancement is the separation of overlapping bands, as shown in Figure 7.6. This shows an experimental spectrum where one band is obscured by the tail of another. The fourth derivative shows the two peaks well resolved.

7.4.5 REPEATED SMOOTHING AND DIFFERENTIATION

When smoothing or differentiating twice it is not necessary to perform two passes of a convolution function, although it may sometimes be practically

---

[*] Since the second derivative of a quadratic polynomial is constant, the baseline should be flat in this case.

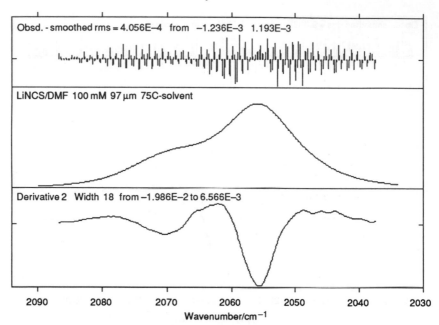

Figure 7.6. Resolution enhancement by numerical differentiation. Upper box: residuals obtained after smoothing. Middle box: experimental data. Lower box: second derivative of the smoothing polynomials

convenient. An equivalent result is obtained by combining the two sets of coefficients together. Let us write down typical expressions for the two passes:

$$y_i^c = \sum_{j=-p,p} C_j y_{i+j}$$

$$y_i^{cc} = \sum_{k=-q,q} D_k y_{i+k}^c$$

$$= \Sigma\Sigma \, C_j D_k y_{i+j+k}$$

We collect together the terms for which $j + k$ is constant:

$$y_i^{cc} = \sum_{h=-p-q,p+q} E_h y_{i+h}$$

$$E_h = \Sigma \, C_j D_k, \qquad j = -p, p; \; k = -q, q; \; j + k = h$$

If the widths of the two functions C and D are m and m' respectively, the width of the combined function is $m + m' - 1$. (Incidentally, two passes of the same function always give an equivalent function with an odd number of points, so that individual functions with an even number of points can be used.)

To illustrate how a second derivative function may be derived from two passes of a first-derivative function, let the latter be the five-point cubic function

$$\mathbf{C} = \mathbf{D} = (1 \quad -8 \quad 0 \quad 8 \quad -1)/12$$

We draw up a table of all the pairwise products of $\mathbf{C}$ with $\mathbf{D}$ (each multiplied by the normalizing constant), then sum the diagonals indicated and finally divide by the product of the normalizing constants

		-2	-1	0	1	2	j
k	D_k	1	-8	0	8	-1	C_j
-2	1	1	-8	0	8	-1	
-1	-8	-8	64	0	-64	8	
0	0	0	0	0	0	0	
1	8	8	-64	0	64	-8	
2	-1	-1	8	0	-8	1	

$$S_{-4} = C_{-2}D_{-2} = 1$$
$$S_{-3} = C_{-2}D_{-1} + C_{-1}D_{-2} = -16$$
$$S_{-2} = C_{-2}D_0 + C_{-1}D_{-1} + C_0D_{-2} = 64$$
$$S_{-1} = C_{-2}D_1 + C_{-1}D_0 + C_0D_{-1} + C_1D_{-2} = 16$$
$$S_0 = C_{-2}D_2 + C_{-1}D_1 + C_0D_0 + C_1D_{-1} + C_2D_{-2} = -130$$
$$\mathbf{E} = (1 \quad -16 \quad 64 \quad 16 \quad -130 \quad 16 \quad 64 \quad -16 \quad 1)/144$$

This nine-point function is equivalent to two passes of the five-point first-derivative function, and is therefore a second-derivative function.

With regard to noise reduction, an expression can be derived in the case of convolution with two passes of a degree 0/1 polynomial. The convolution coefficients for 1 pass of an m-point linear smoothing function take the simple form of $1/m$. The convolution function equivalent to two passes of the same m-point function has coefficients given by k/m^2 ($k = 1 - m, m - 1$). It follows that the variance of the noise remaining after two passes is given by

$$\frac{2m^2 + 1}{3m^3} \sigma^2$$

which approximates to $(2/3m)\,\sigma^2$ for large m. The variance of the noise remaining after one pass is $(1/m)\,\sigma^2$. On the second pass the variance is reduced by a factor of $\frac{2}{3}$; the noise level (standard deviation) is therefore

Polynomials

reduced to about 80% of the level remaining after one pass. O'Haver and Begley (1981) have published some formulae for derivatives for this case.

For polynomials of degree 2/3 the noise remaining after repeated passes of a smoothing function can be obtained numerically. The convolution function equivalent to $p + 1$ passes is built up recursively by convolution of the p-pass function with the original smoothing function. The noise remaining is given by the sum of squared p-pass convolution coefficients. Some results for smoothing by quadratic/cubic polynomials are shown in Figure 7.7. It can be seen that the first few passes give a useful reduction in the noise remaining. For example, with 13 passes of a five-point function the amount of noise remaining is virtually halved as compared to a single pass and with 11 passes of a 21-point function the variance of the noise remaining is a mere 5% of the

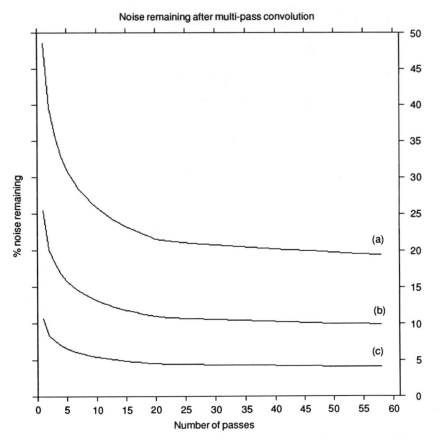

Figure 7.7. Variance of the noise remaining for multipassing with a quadratic/cubic smoothing function, as a percentage of the original noise variance: (a) five-point function, (b) nine-point function, (c) 21-point function

original variance. It would seem therefore that multipassing offers a means of substantially improving the signal-to-noise ratio in data.

There is one drawback—the effective width of the multipass smoothing function. For p passes of an m-point function the equivalent single-pass width is $p(m-1)+1$. Therefore, 11 passes of a 21-point function are equivalent to one pass of a 221-point function, with consequent loss of information at 110 points at each end of the data. To overcome this problem either there must be plenty of redundant data or one of the techniques for end-smoothing described in Section 7.4.1 must be used. The latter was the approach used by Procter and Sherwood (1980) to smooth X-ray photoelectron spectra. They found that some 'end-effects' still remained, but that 40–80 passes of the smoothing function gave reliable results on data with a signal-to-noise ratio as low as 5:1!

Another possible drawback to massive multipassing is that distortion is cumulative, so multipassing with an optimal smoothing function will introduce unacceptable distortion. The remedy in this case is to multipass with a narrower function.

Finally, let us anticipate the discussion of Section 9.5 and consider the multipass smoothing function in the Fourier codomain. It may be shown that the multipass function is more effective at removing both low- and high-frequency noise, the latter being virtually eliminated.

7.4.6 EFFECT OF CONVOLUTION ON CORRELATION

One thing that is not often recognized is that convolution changes the correlation of errors. To see this let us first take a simple example. Suppose that the variance–covariance matrix of the observations is a unit matrix multiplied by a constant σ^2, equivalent to postulating that the data errors are equal and uncorrelated. By the law of propagation of errors (Section 2.2) if $\mathbf{y}^c = \mathbf{Ty}$,

$$\mathbf{COV}(\mathbf{y}^c) = \mathbf{T}\ \mathbf{COV}(\mathbf{y})\ \mathbf{T}^T$$
$$= \sigma^2 \mathbf{TT}^T$$

Rather than give an algebraic expression for the covariance let us write down the numbers involved for a five-point quadratic/cubic smooth:

$$
\begin{bmatrix} y_3^c \\ y_4^c \\ y_5^c \\ y_6^c \\ y_7^c \\ y_8^c \\ \cdots \end{bmatrix}
= \frac{1}{35}
\begin{bmatrix}
-3 & 12 & 17 & 12 & -3 & 0 & 0 & 0 & 0 & 0 & \cdots \\
0 & -3 & 12 & 17 & 12 & -3 & 0 & 0 & 0 & 0 & \cdots \\
0 & 0 & -3 & 12 & 17 & 12 & -3 & 0 & 0 & 0 & \cdots \\
0 & 0 & 0 & -3 & 12 & 17 & 12 & -3 & 0 & 0 & \cdots \\
0 & 0 & 0 & 0 & -3 & 12 & 17 & 12 & -3 & 0 & \cdots \\
0 & 0 & 0 & 0 & 0 & -3 & 12 & 17 & 12 & -3 & \cdots \\
\cdots & \cdots & \cdots & \cdots & \cdots & \cdots & \cdots & \cdots & \cdots & \cdots & \cdots
\end{bmatrix}
\begin{bmatrix} y_1 \\ y_2 \\ y_3 \\ y_4 \\ y_5 \\ y_6 \\ \cdots \end{bmatrix}
$$

The covariances are obtained by summing products of rows. For instance,

$$\text{COV}(y_3^\varsigma, y_4^\varsigma) = (12 \times -3 + 17 \times 12 + 12 \times 17 - 3 \times 12)/35^2\sigma^2$$

The form of the lower triangle of the matrix of correlation coefficients for this example is as follows:

y_4^ς	0.56						
y_5^ς	0.07	0.56					
y_6^ς	-0.12	0.07	0.56				
y_7^ς	0.02	-0.12	0.07	0.56			
y_8^ς	0	0.02	-0.12	0.07	0.56		
y_9^ς	0	0	0.02	-0.12	0.07	0.56	
y_{10}^ς	0	0	0	0.02	-0.12	0.07	0.56
...	...	...		...	...	...	... ...
	y_3^ς	y_4^ς	y_5^ς	y_6^ς	y_7^ς	y_8^ς y_9^ς ...

This shows that there is a correlation coefficient of 0.56 between adjacent points, 0.07 between points 2 apart, -0.12 between points 3 apart and 0.02 between points 4 apart. Points separated by 5 or more have zero correlation coefficients. The general picture can be seen by extrapolating from this example. Even though the original data errors were uncorrelated, the errors in the smoothed data are correlated. The correlation extends $m-1$ points on either side of a calculated point, except at the ends of the data.

An important consequence of the introduction of correlation is that it affects any subsequent operation that might be performed on the data. For example, it might be felt to be worth while to smooth data before performing a curve resolution. This is a dangerous procedure for two reasons. Firstly, as we have already seen, smoothing inevitably introduces some distortion into the data. Secondly, because of the introduction of correlation, even when the original data were uncorrelated the normal unweighted or diagonally weighted least-squares calculation is inappropriate.

7.4.7   PRECISE LOCATION OF MAXIMA, MINIMA AND INFLECTION POINTS

By definition, the positions of maxima and minima are the places at which the odd derivatives are zero and inflection points are located where the even derivatives are zero. The least-squares method may be used to locate these zeros in noisy data.

The simplest procedure for locating a maximum or minimum is to calculate an odd derivative, e.g. the first derivative, and locate the two points on either side of the extremum. Designating these two points as x_1, d_1 and x_2, d_2, the zero may be found by linear interpolation:

$$x_{d=0} = x_1 - d_1 \frac{x_2 - x_1}{d_2 - d_1}$$

Consider next the case of fitting a parabola $y = a_0 + a_1 z + a_2 z^2$ to equally spaced points in the neighbourhood of an extremum. The position of the extremum is given by

$$z_{d=0} = -a_1/2a_2; \qquad z = (x - \bar{x})/h$$

Therefore, if the polynomial coefficients a_1 and a_2 are known, the position of the maxima or minima that occur between data points may be estimated, and so may the error in that position. For example, if we fit a quadratic function to five equally spaced points, the position is given by

$$x_{d=0} = x_i + h \frac{14y_{i-2} + 7y_{i-1} \qquad -7y_{i+1} - 14y_{i+2}}{10y_{i-2} - 5y_{i-1} - 10y_i - 5y_{i+1} + 10y_{i+2}}$$

Formulae such as this work well when the points spanned by the polynomial are all close to the extremum. Johnson and Harmony (1973) have proposed a correction formula that can be applied when the data are relatively sparse around maxima in Gaussian or Lorentzian peaks, and have applied it to microwave spectral data.

An inflection point can be located only with a polynomial of degree 3 or more. The easiest procedure is to use a cubic polynomial, so that the position of the inflection point is given by

$$z_{d=0} = \frac{-a_2}{3a_3}$$

Likewise, the position of a zero in the third derivative is most easily estimated using a quartic polynomial fit:

$$z_{d=0} = \frac{-a_3}{4a_4}$$

The calculation of the error on $x_{d=0}$ is more complicated as the errors on the parameters are needed. Here is a procedure of general application. First set up the normal equations for a polynomial of degree $n - 1$:

$$y_i^{calc} = a_0 + a_1 x_1 + \cdots + a_n x_i^{n-1}$$

for m points $(x_k \cdots x_{k+m-1})$ in the region of the extremum $(m > n)$. Then solve for the n parameters and locate the position of the extremum. Next complete the inversion of the normal equations matrix and calculate the sum of squared residuals as

$$S = \sum_{j=k, k+m-1} (y_i^{obs} - a_0 - a_1 x_j - a_2 x_j^2 - \cdots)^2$$

and obtain the errors on the parameters by the standard method (Section 3.6).

An interesting application is the location of maxima that are obscured by overlap by other functions, as in the data shown in Figure 7.6. If the peak maximum cannot be seen in the original data there will be no place at which the first derivative, i.e. slope, will be zero. The third derivative, shown in Figure 7.8, clearly shows the two zeros associated with the two underlying maxima. Note that whereas a first derivative goes from positive to negative in the region of a maximum, the third derivative goes from negative to positive.

Location of maxima and minima is not restricted to the situation in which the data points are equally spaced, but can be done in the general case by direct fitting of an appropriate polynomial. When the data are equally spaced convolution coefficients may be used, but the results must be corrected for the parameter scaling that is implied in the use of these coefficients.

7.4.8 INTEGRATION

Although there is a vast literature on the integration of functions, little seems to have been published on the integration of noisy data. The reason for this is probably that there is little call to perform this operation. In most cases

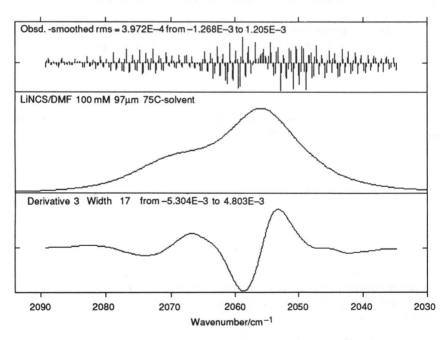

Figure 7.8. Use of the third derivative to locate maxima. Upper box: residuals obtained after smoothing. Middle box: experimental data. Lower box: third derivative of the smoothing polynomials

where the area under a curve is significant there is a simple relationship between the area and another parameter such as the maximum height of the curve. This is certainly true of Gaussian and Lorentzian bands.

Nevertheless, the method of smoothing by piecewise fitting of a polynomial can easily be extended to integration. If we write the polynomial as

$$y = a_0 + a_1 x + a_2 x^2 + \cdots$$

we may also write a definite integral as

$$\int_{-p}^{p} y \, dx = [a_0 x + a_1 x^2/2 + a_2 x^3/3 + \cdots]_{-p}^{p}$$
$$= 2 a_0 p + \tfrac{2}{3} a_2 p^3 + \cdots$$

and hence deduce a formula similar to the convolution formula for calculating the area. There is one major difference in the procedure: the area is calculated for the first m points, then for the next m points, and so forth, and the individual areas are added together. Fitting a quadratic to three data points results in Simpson's one-third rule and fitting a cubic to four data points results in the three-eighths rule. The coefficients in Table 7.2 are for integration with

Table 7.2. Coefficients for integration over m points with fitting by quadratic/cubic polynomial

				Number of points					
3	5	7	9	11	13	15	17	19	21
									77 140
								1683	82 840
							2096	1836	87 940
						9 191	2336	1971	92 440
					660	10 556	2544	2088	96 340
				795	792	11 711	2720	2187	99 640
			392	1020	900	12 656	2864	2268	102 340
		7	560	1195	984	13 391	2976	2331	104 440
	44	12	680	1320	1044	13 916	3056	2376	105 940
1	104	15	752	1395	1080	14 231	3104	2403	106 840
4	124	16	776	1420	1092	14 336	3120	2412	107 140
1	104	15	752	1395	1080	14 231	3104	2403	106 840
	44	12	680	1320	1044	13 916	3056	2376	105 940
		7	560	1195	984	13 391	2976	2331	104 440
			392	1020	900	12 656	2864	2268	102 340
				795	792	11 711	2720	2187	99 640
					660	10 556	2544	2088	96 340
						9 191	2336	1971	92 440
							2096	1836	87 940
3	105	14	693	1287	1001	13 260	2907	2261	100 947

Normalizing factor

least-squares fitting of parabolas. The seven-point formula looks particularly attractive $(h = x_{i+1} - x_i)$:

$$I = h[7(y_{i-3} + y_{i+3}) + 12(y_{i-2} + y_{i+2}) + 15(y_{i-1} + y_{i+1}) + 16y_i]/14$$

7.5

7.5.1 FITTING A PARABOLA

The model for a parabola in its simplest form is

$$y = a_0 + a_1 x + a_2 x^2$$

The three normal equations in the unknown parameters a_0, a_1 and a_2 can be solved directly by the general method, but this usually requires a computer and a program. As there will be instances when one would like to fit a parabola with the aid of a hand-calculator, we shall now look at ways in which this can be done. When the data points are equally spaced the methods of Section 7.4, which involve the solution of single or pairs of normal equations, may be employed. Indeed, when the data are equally spaced, cubic models can be fitted easily by hand either by the direct approach or by the method of orthogonal polynomials.

When the data points are not equally spaced there is inevitably a large amount of calculation involved. Three methods will now be outlined. Each method is expressed as though unit weights are used, but weights can be introduced as described in Section 3.3.2.

7.5.2 DIRECT INVERSION OF THE NORMAL EQUATIONS MATRIX AND MULTIPLYING OUT

We first write down the normal equations matrix and then invert it:

$$\mathbf{J}^T\mathbf{J} = \begin{bmatrix} m & \Sigma x & \Sigma x^2 \\ \Sigma x & \Sigma x^2 & \Sigma x^3 \\ \Sigma x^2 & \Sigma x^3 & \Sigma x^4 \end{bmatrix} = \begin{bmatrix} m & s_1 & s_2 \\ s_1 & s_2 & s_3 \\ s_2 & s_3 & s_4 \end{bmatrix}$$

$$(\mathbf{J}^T\mathbf{J})^* = \begin{bmatrix} s_2 s_4 - s_3 s_3 & s_2 s_3 - s_1 s_4 & s_1 s_3 - s_2 s_2 \\ s_2 s_3 - s_1 s_4 & m s_4 - s_2 s_2 & s_1 s_2 - m s_3 \\ s_1 s_3 - s_2 s_2 & s_1 s_2 - m s_3 & m s_2 - s_1 s_1 \end{bmatrix}$$

$(\mathbf{J}^T\mathbf{J})^*$ is known as the adjoint matrix of $\mathbf{J}^T\mathbf{J}$, and it is equal to the inverse $(\mathbf{J}^T\mathbf{J})^{-1}$ multiplied by the determinant of $\mathbf{J}^T\mathbf{J}$. It follows that the determinant is given by

$$\det = m(s_2 s_4 - s_3 s_3) + s_1(s_2 s_3 - s_1 s_4) + s_2(s_1 s_3 - s_2 s_2)$$

and the parameters are given by

$$a_0 = [(s_2s_4 - s_3s_3)\Sigma y + (s_2s_3 - s_1s_4)\Sigma xy + (s_1s_3 - s_2s_2)\Sigma x^2 y]/\det$$
$$a_1 = [(s_2s_3 - s_1s_4)\Sigma y + (ms_4 - s_2s_2)\Sigma xy + (s_1s_2 - ms_3)\Sigma x^2 y]/\det$$
$$a_2 = [(s_1s_3 - s_2s_2)\Sigma y + (s_1s_2 - ms_3)\Sigma xy + (ms_2 - s_1s_1)\Sigma x^2 y]/\det$$

The advantage of this procedure is that the errors on the parameters can be easily obtained from the adjoint matrix.

7.5.3   REDUCING THE ORDER

The first normal equation can be written out in full and solved for the constant parameter:

$$ma_0 + \Sigma xa_1 + \Sigma x^2 a_2 = \Sigma y$$
$$a_0 = (\Sigma y - \Sigma xa_1 - \Sigma x^2 a_2)/m$$
$$= \bar{y} - \bar{x}a_1 - \bar{x^2} a_2$$
$$\bar{y} = \Sigma y/m, \qquad \bar{x} = \Sigma x/m, \qquad \bar{x^2} = \Sigma x^2/m$$

This expression can be substituted into the other normal equations to yield a set of two equations in the remaining two parameters. After some manipulation we may obtain the two equations in the following form:

$$\Sigma (x - \bar{x})^2 a_1 + \Sigma (x - \bar{x})(x^2 - \bar{x^2})a_2 = \Sigma (y - \bar{y})(x - \bar{x})$$
$$\Sigma (x - \bar{x})(x^2 - \bar{x^2})a_1 \quad + \Sigma (x^2 - \bar{x^2})^2 a_2 = \Sigma (y - \bar{y})(x^2 - \bar{x^2})$$

After solving these two equations in the usual way, the first parameter is obtained from the other two.

7.5.4   USE OF ORTHOGONAL POLYNOMIALS

To use orthogonal polynomials we first rewrite the model as

$$y = b_0 + b_1(x + \beta) + b_2(x^2 + \gamma x + \delta)$$
$$[J_i] = [1 \; (x_i + \beta) \; (x_i^2 + \gamma x_i + \delta)]$$

and derive the normal equations from this model (Section 7.3):

$$\mathbf{J^T J} = \begin{bmatrix} m & \Sigma (x + \beta) & \Sigma (x^2 + \gamma x + \delta) \\ \Sigma (x + \beta) & \Sigma (x + \beta)^2 & \Sigma (x + \beta)(x^2 + \gamma x + \delta) \\ \Sigma (x^2 + \gamma x + \delta) & \Sigma (x + \beta)(x^2 + \gamma x + \delta) & \Sigma (x^2 + \gamma x + \delta)^2 \end{bmatrix}$$

$$\mathbf{J^T y} = \begin{bmatrix} \Sigma y \\ \Sigma (x + \beta)y \\ \Sigma (x^2 + \gamma x + \delta)y \end{bmatrix}$$

If the polynomials (1), $(x + \beta)$ and $(x^2 + \gamma x + \delta)$ are mutually orthogonal over the set of data points, all the off-diagonal terms in the normal equations matrix must, by definition, be equal to zero. We can work through the consequences of this to find the parameters one by one:

First equation: $\quad m b_0 = \Sigma\, y \qquad b_0 = \bar{y}$

$$\mathbf{J}^T\mathbf{J}_{12} = 0 \qquad\quad \Sigma\, (x + \beta) = 0 \qquad\quad \beta = -\bar{x}$$

Second equation: $\quad \Sigma\, (x + \beta)^2 b_1 = \Sigma\, (x + \beta)y \qquad b_1 = \dfrac{\Sigma\, (x - \bar{x})y}{\Sigma\, (x - \bar{x})^2}$

$$\mathbf{J}^T\mathbf{J}_{13} = 0 \qquad \Sigma\, (x^2 + \gamma x + \delta) = 0 \qquad \delta = -\bar{x}^2 - \gamma\bar{x}$$

$$\mathbf{J}^T\mathbf{J}_{23} = 0 \qquad \Sigma\, (x - \bar{x})(x^2 + \gamma x + \delta) = 0 \qquad \gamma = \dfrac{-\Sigma\, (x - \bar{x})(x^2 - \bar{x}^2)}{\Sigma\, (x - \bar{x})^2}$$

Third equation: $\quad b_2 = \dfrac{\Sigma\, (x^2 + \gamma x + \delta)y}{\Sigma\, (x^2 + \gamma x + \delta)^2}$

Finally, we can obtain the usual parameters by expanding the model and equating coefficients:

$$a_0 = b_0 + b_1\beta + b_2\delta$$
$$a_1 = b_1 + b_2\gamma$$
$$a_2 = b_2$$

Although the procedure is fairly easy to implement, it suffers from the disadvantage that the errors and correlation coefficients of the a parameters are more difficult to obtain (see Guest, 1961, for more details).

7.6

7.6.1   THE STRAIGHT LINE

It may seem odd to the reader that this section has been left until the end of the chapter on polynomial fitting. Fitting a straight line is at the same time the simplest of all least-squares procedures, and the one about which the most has been written. There is even a whole book devoted to it, by Acton (1959). However, almost all the 'complications' and 'special cases' described in the elementary texts can be seen as specific examples of the general procedures as described in Chapter 3. Rather, some of the methods should be viewed as simplifications of the general procedure. Therefore in this section we will concentrate on some of the consequences of those simplifications.

The general statement of the normal equations, $\mathbf{J}^T\mathbf{W}\mathbf{J}\mathbf{p} = \mathbf{J}^T\mathbf{W}\mathbf{y}$, when applied to the linear model

$$y_i^{\text{calc}} = a_0 + a_1 x_i$$

results in the two normal equations

$$(\Sigma\Sigma\, W_{ik})a_0 + (\Sigma\Sigma\, W_{ik}x_k)a_1 = \Sigma\Sigma\, W_{ki}y_i$$
$$(\Sigma\Sigma\, W_{ik}x_k)a_0 + (\Sigma\Sigma\, W_{ik}x_k x_i)a_1 = \Sigma\Sigma\, W_{ki}x_k y_i$$

$$\Sigma\Sigma\cdots \equiv \sum_{i=1,m}\ \sum_{k=1,m}\cdots$$

In the general case the weight matrix $\mathbf{W}$ is the inverse of the variance–covariance matrix which, when there are errors on both x and y, is obtained by combining together the variance–covariance matrices on x and y, as discussed in Section 3.4. The general solution is given by

$$a_1 = \frac{-(\Sigma\Sigma\, W_{ik}x_k)(\Sigma\Sigma\, W_{ki}y_i) + (\Sigma\Sigma\, W_{ik})(\Sigma\Sigma\, W_{ki}x_k y_i)}{(\Sigma\Sigma\, W_{ik})(\Sigma\Sigma\, W_{ik}x_k x_i) - (\Sigma\Sigma\, W_{ik}x_k)^2}$$

$$a_0 = \frac{(\Sigma\Sigma\, W_{ki}y_i) - (\Sigma\Sigma\, W_{ik}x_k)a_1}{\Sigma\Sigma\, W_{ik}}$$

We now introduce the concept of the globally weighted means, $\bar{x}$ and $\bar{y}$:

$$\bar{x} = \frac{\Sigma\Sigma\, W_{ik}x_i}{\Sigma\Sigma\, W_{ik}}$$

$$\bar{y} = \frac{\Sigma\Sigma\, W_{ik}y_i}{\Sigma\Sigma\, W_{ik}}$$

In terms of these means the constant parameter takes the simple form

$$a_0 = \bar{y} - a_1\bar{x}$$

and it can be 'eliminated' from the model, so that the model can be written in the alternative form

$$y - \bar{y} = a_1(x - \bar{x})$$

This equation for the model reveals an important property of the least-squares straight line, namely that it passes through the global mean point $(\bar{x}, \bar{y})$, i.e. when $x = \bar{x}$, $y = \bar{y}$. It also shows that the slope of the line is independent of the origin of the x and y coordinates, which is intuitively obvious. The slope parameter can also be expressed in terms of the global means:

$$a_1 = \frac{\Sigma\Sigma\, W_{ik}(x_i - \bar{x})(y_k - \bar{y})}{\Sigma\Sigma\, W_{ik}(x_i - \bar{x})(x_k - \bar{x})}$$

We may simplify and utilize these expressions in various ways. Let us begin by defining five new subsidiary quantities:

$$sw = \Sigma\Sigma\, W_{ik}$$

$$sxx = \Sigma\Sigma\, W_{ik}x_ix_k$$

$$cxx = \Sigma\Sigma\, W_{ik}(x_i - \bar{x})(x_k - \bar{x}) = \Sigma\Sigma\, W_{ik}x_ix_k - (\Sigma\Sigma\, W_{ik}x_i)^2/\Sigma\Sigma\, W_{ik}$$

$$cxy = \Sigma\Sigma\, W_{ik}(x_i - \bar{x})(y_k - \bar{y})$$
$$= \Sigma\Sigma\, W_{ik}x_iy_k - (\Sigma\Sigma\, W_{ik}x_i)\Sigma\Sigma\, W_{ik}y_i)/\Sigma\Sigma\, W_{ik}$$

$$cyy = \Sigma\Sigma\, W_{ik}(y_i - \bar{y})(y_k - \bar{y}) = \Sigma\Sigma\, W_{ik}y_iy_k - (\Sigma\Sigma\, W_{ik}y_i)^2/\Sigma\Sigma\, W_{ik}$$

We may now write the two parameters in terms of these expressions and the global means:

$$a_0 = \bar{y} - a_1\bar{x}$$

$$a_1 = cxy/cxx$$

The weighted sum of squared residuals can also be expressed quite simply:

$$S = \Sigma\Sigma\, W_{ik}(y_i - a_0 - a_1x_i)(y_k - a_0 - a_1x_k)$$
$$= \Sigma\Sigma\, W_{ik}[y_i - \bar{y} - a_1(x_i - \bar{x})]\,[y_k - \bar{y} - a_1(x_k - \bar{x})]$$
$$= cyy - a_1cxy$$

This enables us to give expressions for the standard deviations and correlation coefficient on the parameters. The whole calculation may be summarized:

$$a_0 = \bar{y} - a_1\bar{x}$$

$$a_1 = cxy/cxx$$

$$S = cyy - a_1cxy$$

$$\sigma_{a_0} = \left(\frac{sxx\ S}{(m-2)sw\ cxx}\right)^{1/2}$$

$$\sigma_{a_1} = \left(\frac{S}{(m-2)sxx}\right)^{1/2}$$

$$\rho = \frac{-sw\ \bar{x}}{(sw\ sxx)^{1/2}}$$

$$r_i = y_i^{\text{obs}} - a_0 - a_1x_i$$

We can also calculate another correlation coefficient. This one shows the overall extent of correlation in the experimental data:

$$\rho_{\text{exp}} = \frac{cxy}{(cxx\ cyy)^{1/2}}$$

Because there are only two normal equations for fitting a straight line, all the quantities of interest can be written down in analytical form; it must be stressed that no least-squares fit is complete without a calculation and plot of the residuals.

The expressions given above are in the most general form: i.e. the weight matrix is not assumed to have any elements equal to zero. Commonly the weight matrix is taken to be diagonal, and this results in some simplification of all the expressions. When the weight matrix is diagonal ($W_{ik} = 0$ when $i \neq k$) the double summations collapse to single summations:

$$\bar{x} = \frac{\Sigma W_{ii} x_i}{\Sigma W_{ii}}, \qquad \Sigma \equiv \sum_{i=1,m}$$

$$\bar{y} = \frac{\Sigma W_{ii} y_i}{\Sigma W_{ii}}$$

$$sw = \Sigma W_i$$

$$sxx = \Sigma W_i x_i^2$$

$$cxx = \Sigma W_{ii}(x_i - \bar{x})^2 = \Sigma W_{ii} x_i^2 - (\Sigma W_{ii} x_i)^2 / \Sigma W_{ii}$$

$$cxy = \Sigma W_{ii}(x_i - \bar{x})(y_i - \bar{y}) = \Sigma W_{ii} x_i y_i - (\Sigma W_{ii} x_i)(\Sigma W_{ii} y_i) / \Sigma W_{ii}$$

$$cyy = \Sigma W_{ii}(y_i - \bar{y})^2 = \Sigma W_{ii} y_i^2 - (\Sigma W_{ii} y_i)^2 / \Sigma W_{ii}$$

It should be noted that the expressions for the parameters, etc., given in the box above, are not changed; it is only the summation ranges of the quantities in those expressions that are different. Moreover, all the required quantities can be calculated using only the weighted sums of x, y, xy, x^2 and y^2, thus making it possible to calculate them 'instantaneously' as each x, y pair is entered. A useful by-product of this is that an x, y point may be added or subtracted to the data set without having to do a complete recalculation.

If the weight matrix is diagonal and all the weights are equal to unity the formulae become even more simple:

$$\bar{x} = \frac{\Sigma x}{m}$$

$$\bar{y} = \frac{\Sigma y}{m}$$

$$sw = m$$

$$sxx = \Sigma x^2$$

$$cxx = \Sigma x^2 - (\Sigma x)^2 / m$$

$$cxy = \Sigma xy - (\Sigma x \Sigma y) / m$$

$$cyy = \Sigma y^2 - (\Sigma y)^2 / m$$

Given that the algebra of straight line fitting is simple, if a little tedious, we may justifiably ask why so much has been written on the subject. There are many answers. Firstly, fitting the straight line has often been used as an introduction to the least-squares process itself. Secondly, it seems more complicated if the weights are not introduced at the outset as we have done here. Thirdly, the cases where x and/or y are subject to error are often treated individually, whereas we have taken the approach that the only difference between them is the way the weights are obtained. Finally, the application of statistics seems to introduce further endless complications.

It therefore remains only to discuss the last two items in order to close this section. The weight matrix for the general case is calculated according to Section 3.4.2 as

$$\mathbf{M} = \mathbf{M}_y + a_1^2 \mathbf{M}_x; \qquad \mathbf{W} = \mathbf{M}^{-1}$$

where $\mathbf{M}_x$ and $\mathbf{M}_y$ are the variance–covariance matrices on x and y. As we have seen, the slope parameter a_1 may be replaced by an estimate of the slope obtained directly from the data, for example $(y_m - y_1)/(x_m - x_1)$. To be rigorous one should repeat the calculation until the calculated slope, a_1, is not significantly different from the slope used to calculate the weights. Of course the whole process becomes simpler if the variance–covariance matrices are diagonal since then $W_{ii} = 1/M_{ii}$, or if there are negligible errors on x, in which case $\mathbf{M} = \mathbf{M}_y$. If both simplifications are applicable $W_{ii} = 1/M_{y,ii}$.

The so-called Berkson case merits a little further discussion. This is the case when the independent variable x is not so much measured as set to predetermined values.[*] In the example in Section 6.8, the samples were prepared from a stock solution by dilution using pipettes, and the choice of pipettes predetermined the concentration of analyte in each sample.

Berkson did a statistical analysis of the straight line fitting process on these systems. He concluded that in these cases it is permissible to omit the errors on the independent (preset) variable from the calculation. Nevertheless, the residuals reflect the effect of errors in both variables x and y.

We have assumed that the variance–covariance matrix can, in principle, always be obtained, but there is one situation in which this is not so. This is the situation when the errors on y, say, are cumulative; this has been discussed in detail by Mandel (1964). A realistic example can be found in chemical kinetics. The gas-phase decomposition of N_2O_5 is reasonably expected to be a first-order process in which the rate of decomposition is

---

[*] This is bad experimental practice when x has to be adjusted to the predetermined value. For example, it is easier to read a burette than to adjust the contents so that the reading is 0.00; this preset value is therefore likely to be subject to a larger error than a measured value.

proportional to the amount of N_2O_5 present:

$$2N_2O_5 \longrightarrow O_2 + 2N_2O_4$$
$$N_2O_4 \rightleftharpoons 2NO_2$$

Using the gas pressure[*] as a measure of the amount present, and taking the pressure to be p_0 at time $t = 0$,

$$\frac{dp}{dt} = -kp$$

$$\ln(p) = -kt + \ln(p_0)$$

Thus, the logarithm of the pressure should plot as a straight line against time. The reaction rate is sensitive to the temperature of the mixture, so that if the temperature is not strictly controlled the extent of reaction at any time t will be falsified, and this error will be present in *all* subsequent measurements. Errors of this kind, which Mandel describes as process errors as opposed to errors of measurement, are cumulative. Mandel then goes on to show that the effect of cumulative errors can by mitigated if, instead of fitting the logarithm of the pressure as a function of time, one fits increments in the logarithm as a function of increments in time.

This procedure can be criticized on two fundamental counts. Firstly, what Mandel calls process errors are really systematic errors, since they affect all subsequent measurements. It would have been much better to improve the thermostatting of the gas mixture so as to remove this process error than to try to accommodate a systematic error into a least-squares procedure. The second criticism is that the system has been artificially linearized by taking logarithms of the true relationship between the experimentally observed quantities of pressure and time:

$$p = p_0\, e^{-kt}$$

If the error in measured pressure is assumed to be a constant, the error in the logarithm of the pressure is far from constant, and so the unweighted straight line fit is inappropriate. It would be safer to fit the observed quantity, total pressure, as a function of time:

$$p_{\text{total}} = (1.5 + \alpha)p_0 - (0.5 + \alpha)p_0\, e^{-kt}$$

Unfortunately the original data of Daniels and Johnston (1921) is defective since they published times and pressures that had been interpolated from amongst the observed values.

---

[*] The partial pressure of N_2O_5 could be obtained from the total pressure by $p = [p_0(1.5 + \alpha) - p_{\text{total}}]/(0.5 + \alpha)$, where α is the degree of dissociation of N_2O_4 into NO_2 calculated from the equilibrium constant, K: $\alpha = \sqrt{[K/(4p_{\text{total}} + K)]}$. Note that α depends on the total pressure.

7.6.2 STATISTICS

Just as the parameters and their errors can be written down in closed form, so can any statistical expression that may be needed. Details can be found in any statistical text. However, this is an opportune place to put a gloss on some of the terms commonly used.

Fitting a straight line is often described as *regression analysis*. The term goes back to a paper by Francis Galton published in 1886 entitled 'Regression towards mediocrity in hereditary stature'. Nowadays statisticians tend to call any straight line fitting a linear regression analysis. If the dependent variable is y and the independent variable is x, the least-squares straight line is said to represent the regression of y on x.

Multiple linear regression is a term applied to a model of the kind

$$y_1 = a_0 + a_1 x_i + b_1 z_i + \cdots$$

where $x, z, \ldots$ are different independent variables. The technique of multiple linear regression is of limited use in the physical sciences, so it will not be discussed in detail. However, the normal equations for this model also take a simple form:

$$a_0 = \bar{y} - a_1 \bar{x} - b_1 \bar{z} \cdots$$

$$a_1 = sxy/sxx$$

$$b_1 = szy/szz$$

$$\cdots$$

where szy, szz are defined in a similar manner to sxy, sxx, as above.

Other statistical considerations such as the evaluation of confidence limits are discussed in Chapter 6.

7.6.3 STRAIGHT LINE THROUGH THE ORIGIN

One must question the wisdom of forcing a straight line to pass through the origin whenever one is dealing with experimental data. Even though theory may suggest that the line should go through the origin, there is no guarantee that it will do so in the presence of experimental error. To force the line onto the origin is to introduce systematic error into an empirical model.

Having said that, the algebra is particularly simple. The model is

$$y_i^{\text{calc}} = a_1 x_i$$

and there is only one normal equation. The solution is

$$a_1 = \frac{sxy}{sxx}$$

$$S = syy - a_1 sxy$$

$$\sigma_{a_1} = \left(\frac{S}{(m-1)sxx}\right)^{1/2}$$

As with the general straight line, different expressions for the quantities in these formulae can be written down, depending on the nature of the weight matrix:

Full weights	Diagonal weights	Unit weights
$sxx = \Sigma\Sigma\, W_{ik} x_k x_i$	$sxx = \Sigma\, W_{ii} x_i^2$	$sxx = \Sigma\, x_i^2$
$sxy = \Sigma\Sigma\, W_{ik} x_k y_i$	$sxy = \Sigma\, W_{ii} x_i y_i$	$sxy = \Sigma\, x_i y_i$
$syy = \Sigma\Sigma\, W_{ik} y_k y_i$	$syy = \Sigma\, W_{ii} y_i^2$	$syy = \Sigma\, y_i^2$

7.6.4   TRANSFORMATION TO LINEAR MODELS

There are a number of model types that can be algebraically transformed into linear ones. We have already met one of these, the exponential model:

$$y = p\, e^{qx}; \qquad \log(y) = \log(p) + qx$$

Many others could be listed. The rationale for performing this transformation is that the straight line is easy to fit and easy to test for applicability to the data. When computers are readily available these reasons are no longer important.

Transformation to linear models is to be strongly discouraged, except for the purpose of visual inspection of the data. This is because the errors on the transformed variables do not have a linear relationship to the errors on the observations. When the fractional errors are small a linear approximation is not too bad. In the case of the exponential, let us assume all errors on y to be equal to σ and that $\sigma/y \ll y$. The error on the logarithm is given approximately by σ/y. Therefore, if you must transform to a linear model, the least you should do is try to transform the experimental errors.

Neglect of the effect on errors of the non-linear transformation can lead to some seriously erroneous conclusions. The work of Cumming *et al.* (1972) provides an example. The pK of a weak acid was found to be a whole unit in error when the correct weighting scheme was not used. Another pernicious example comes from the field of enzyme kinetics. Some systems follow the Michaelis–Menten law:

$$k = \frac{V_{max}x}{K_M + x}$$

k is an observed rate constant, often obtained from the initial slope of substrate concentration as a function of time. V_{max} and K_M are the parameters and x is the concentration of a second substrate. The equation can be transformed into linear form:

$$\frac{1}{k} = \frac{1}{V_{max}} + \frac{K_M/V_{max}}{x}$$

$1/k$ is then plotted against $1/x$. What makes this transformation so undesirable is the fact that the rate constants k are subject to relatively large errors; indeed, 'outliers' are not uncommon. It is preferable to fit the Michaelis–Menten expression directly and apply the usual criteria to decide whether the law is applicable.

8

Fitting Functions

8.1 SPLINE FUNCTIONS

Spline functions are very useful when there is no analytical model for the data.
They provide a general and versatile means for finding a smooth approximation to the data which can be used in various ways. In this section we shall
concentrate on one form of spline function, the form that lends itself to fitting
by the least-squares method.

What is a spline function? It is a function of an independent variable, x,
such that at any one point its value is given by a polynomial in x. However,
the polynomial at one point, x_i, is not necessarily the same as the polynomial
at another point, x_j. The places where the polynomial changes from one form
to another are termed *knots, nodes* or *breakpoints*. The most important
property of the spline function is that it is a *continuous* function; one or more
of its derivatives may also be continuous. To illustrate this, consider the
following function:

$$y = x^2/2 \qquad\qquad \text{for } 0 \leqslant x < 1$$
$$y = (-2x^2 + 6x - 3)/2 \qquad \text{for } 1 \leqslant x < 2$$
$$y = (3 - x)^2/2 \qquad\qquad \text{for } 2 \leqslant x < 3$$

This function is shown in Figure 8.1. The breakpoints are given by $x = 0$, 1,
2 and 3. This function does not exist when x is less than 0 or greater than or
equal to 3. It is therefore *discontinuous* outside the region of its definition.
Within the region of definition it is continuous, as can clearly be seen from the
graph. In particular, adjacent portions of the function have the same values
at the breakpoints:

$$x^2/2 = (-2x^2 + 6x - 3)/2 = \tfrac{1}{2} \qquad \text{at } x = 1$$
$$(-2x^2 + 6x - 3)/2 = (3 - x)^2/2 = \tfrac{1}{2} \qquad \text{at } x = 2$$

Fitting Functions

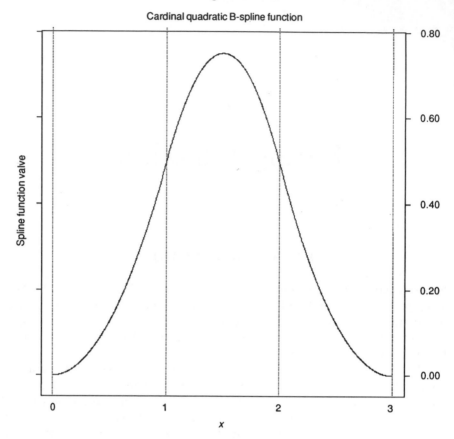

Figure 8.1. The cardinal B-spline of degree 2 on knots at $x = 0$, 1, 2 and 3

Furthermore, the first derivative is also continuous:

$$x = 3 - 2x = 1 \qquad \text{at } x = 1$$
$$3 - 2x = -(3 - x) = -1 \qquad \text{at } x = 2$$

The second derivative is discontinuous at both $x = 1$ and $x = 2$.

This function illustrates the essential characteristics of a spline function. It is a function both of the independent variable and of the positions of the breakpoints. It consists of adjoining pieces and can be said to be *piecewise polynomial*. Each piece is a polynomial of degree n; the spline function is said to be of *order $n + 1$*. The function (and perhaps some of its derivatives) are continuous, but the nth derivative (and perhaps others) is discontinuous at the breakpoints.

The construction of spline functions for least-squares fitting proceeds by linear combination of a set of *basis splines*. The basis splines are denoted by

$B_{ik}(x)$, where i is the index of the left-most breakpoint and k is the order of the polynomial.* Basis splines (B-splines) are not the only possible choice for a basis, but we shall not discuss the others here. For a detailed description of the B-splines, their derivation and properties see de Boor (1978).

There is only one B-spline of given degree on a set of breakpoints. If the latter are denoted by $t_1, t_2, \ldots$ all the B-splines that exist in a given interval can be evaluated by means of the following recursion formula:

$$t_i \leqslant x < t_{i+1}$$

$$B_{i,1}(x) = 1$$

$$B_{i,k}(x) = \frac{x - t_i}{t_{i+k-1} - t_i} B_{i,k-1}(x) + \frac{t_{i+k} - x}{t_{i+k} - t_{i+1}} B_{i+1,k-1}(x)$$

This formula gives no indication of where the B-splines came from, but it does provide a very convenient way of calculating them, both numerically and analytically. Indeed, the B-spline of Figure 8.1 was evaluated as follows:

$$
\begin{array}{lll}
k = 1: & B_{i,1} & = 1 \\
 & B_{i+1,1} = 0 \\
k = 2: & B_{i,2} & = x - t_i \\
 & B_{i-1,2} = t_{i+1} - x \\
 & B_{i+1,2} = 0 \\
k = 3: & B_{i,3} & = (x - t_i)^2/2 \\
 & B_{i-1,3} = [(x - t_{i-1})(t_{i+1} - x) + (t_{i+2} - x)(x - t_i)]/2 \\
 & B_{i-2,3} = (t_{i+1} - x)^2/2
\end{array}
$$

These expressions give the values of the polynomial pieces on the interval $t_i \leqslant x < t_{i+1}$, and by setting t_i to 0, 1 and 2 the three pieces of the B-spline can be evaluated. Clearly, the process could be continued with $k = 4, 5, \ldots$.

A number of properties of the B-splines of degree k (order $k + 1$) should be noted:

(a) They have limited *support*, and only exist over a set of $k + 1$ knots.
(b) The knots must be in a strictly non-decreasing sequence, that is $t_{i+1} \geqslant t_i$.
(c) The B-splines on a given set of knots have no adjustable parameters.
(d) They cannot take negative values.
(e) They are continuous over the range of support. In the definitions given here they are continuous from the left.
(f) The $(k - 1)$th derivative is discontinuous at the knots. Other derivatives may be continuous.

---

* Just as the functions $x, x^2, x^3, \ldots$ can be said to be a *basis* for all polynomial functions, in the sense that all polynomials are linear combinations of them, so the basis splines $B_{ik}(x)$ can be linearly combined together to form all possible spline functions on the given set of knots.

The last point needs some amplification. If all the knots are distinct all the derivatives down to the $(k-2)$th are continuous. However, the knots do not need to be distinct. If two knots coincide one condition of continuity in the derivatives disappears. Therefore, when fitting data there must be k coincident knots at the first and last data points since the first spline function is only supported on the interval $t_1 \leqslant x < t_2$, and the last on the interval $t_{m-1} \leqslant x < t_m$. The fact that the derivatives do not need to be continuous could be of great value in fitting certain kinds of experimental data that do not have continuous derivatives.

The derivatives of the B-splines are also easy to evaluate. The first derivative is simply a combination of two B-splines of order $k-1$:

$$\frac{dB_{i,k}(x)}{dx} = (k-1)\frac{-B_{i+1,k-1}(x)}{t_{i+k}-t_{i+1}} + \frac{B_{i,k-1}(x)}{t_{i+k-1}-t_i}$$

From this it follows that spline functions obtained by linear combination of B-splines are also easy to differentiate and integrate.

The general spline function of degree k on a given set of knots can be seen as a linear combination of B-splines:

$$y(x) = \sum_{\text{all } i} \alpha_i B_{i,k}(x)$$

The linear combination has the same conditions of continuity at the breakpoints as the B-splines. Therefore, to fit a spline function to a set of data one must first define the breakpoints and then determine the coefficients α_i by *linear* least squares. The normal equations matrix is banded, with k non-zero entries in each row, and can be factored efficiently by a specially adapted version of Choleski factorization (see de Boor).

The problem in using spline functions is to determine where to place the breakpoints and how many of them to use. The simplest choice is for the breakpoints to be evenly spaced. This implies that there is only one kind of B-spline to be used, the so-called cardinal B-spline. (The function illustrated in Figure 8.1 is the cardinal B-spline of degree 2.) Fitting with the cardinal B-spline is therefore easier than fitting with a general set of B-splines. Unfortunately, fitting with equally spaced knots is not often satisfactory. At the other extreme we can place a breakpoint at each data point. This would provide the interpolating spline function, which passes through all the data points.

De Boor has suggested a number of schemes for deriving knot placements which may be more or less optimal. The idea of an optimal knot placement scheme is that it enables one to reduce the number of knots needed to obtain a satisfactory fit, and by so doing gives a smoother fitting function. Unfortunately none of the suggested schemes is guaranteed to give a good answer.

It is also possible to regard the breakpoint positions as refinable parameters in the least-squares fitting process, though refinement of these parameters turns the linear least-squares process into a non-linear one. For instance, consider trying to fit a spline function to a standard Gaussian of unit height and half-width, with no noise:

$$y^{obs} = e^{-z^2}, \qquad z = x\sqrt{\log_e 2z}; \qquad -5 \leqslant x \leqslant 5$$

This was fitted by a spline function of degree 5 on seven internal knots. The refined positions of these knots were found to be $z = 0$, ± 1.055, ± 2.187. With these knots the fit is remarkably good: the root mean square of the difference between Gaussian and spline functions was found to be about 1/29 000 of the maximum height of the Gaussian (Gans and Gill, 1984).

The extra effort involved in refining the knot positions is rarely justified if there is a scheme that provides reasonably good knot placements. One fairly straightforward criterion for knot placement is that the separation between knots should decrease as the curvature of the data increases. This is equivalent to saying that when the data are strongly curved they can only be represented by a short piece of polynomial, whereas when the curvature is small a polynomial may adequately fit a longer piece of the data.

I was interested in spline function fitting as an alternative to the convolution method for numerical differentiation of spectroscopic curves. We can now make a direct comparison of the two methods. In Figure 8.2 we see the second derivative of the same spectrum as is shown in Figure 5.1. This time the spectrum was fitted with a spline function of degree 5, having 30 internal knots, which are shown as bars on the observed spectrum. (The spacing between the knots was taken as proportional to the square root of the data intensity, which is a rough measure of data curvature.) The quality of the fit is almost identical, as shown by the values of the r.m.s. residuals. This should not surprise us too much, as in both cases we have used polynomial approximations to the data. The other interesting feature of Figure 8.2 is that the calculated second derivative is a spline function of degree 3 on the same set of knots. To obtain this it was necessary to fit the spectrum with a spline function of degree 5. We may conclude that neither the spline function method nor the convolution method shows any marked advantage in this application.

What are the potential applications of spline function fitting? An obvious one is data compression. The data shown in the figure consist of 620 points. The spline function representation consists of 30 knot values and 30 B-spline coefficients, together with a handful of control integers. This represents an approximately ten-fold reduction in the amount of storage needed to represent the spectrum. All that is lost in the process is the experimental noise. Otherwise the spline function is a good representation of the data.

Another potential application is in the decomposition of spectroscopic curves. Suppose that we wish to analyse a mixture consisting of two components

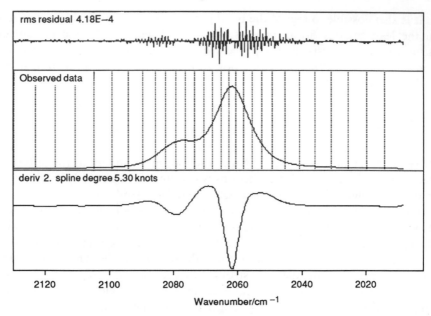

Figure 8.2. Numerical differentiation by the spline function method. Upper box: residuals (data—quintic spline function). Middle box: experimental data. Lower box: second derivative of the spline function. The knot positions are indicated by vertical dotted lines

whose spectra overlap, and suppose that the spectra of the individual components consist of one or more asymmetric bands; the standardized spectra of the components could be represented by spline functions, and the spectrum of a mixture could be decomposed into a sum of the component spectra.

In summary, a spline function can be used to provide an empirical relationship between a set of observations whenever there is no convenient analytical relationship. This includes fitting data that have sharp corners, which are difficult to fit using any other functions.

8.2

8.2.1 EXPONENTIAL FUNCTIONS

The model for a single exponential function may be written as

$$y = c + p\ e^{qx}$$

There is a convenient method of obtaining initial estimates for the parameters. Three data points are required to determine the three parameters. If the x

values of these three points are x_1, x_c and x_3, the centre point must be in the middle of the other two, i.e.

$$x_3 - x_c = x_c - x_1$$

If the data are equally spaced it is convenient to take x_1 as the first point and x_3 as the last (or last but one) point. If the data are not equally spaced the value of y at a point in the centre can be estimated by linear interpolation:

$$y_c = y_{c-}^{obs} + \frac{(x_c - x_{c-})(y_{c+}^{obs} - y_{c-}^{obs})}{x_{c+} - x_{c-}}$$

where x_{c-} and x_{c+} are the x values of the points either side of the central point. Let us designate the three points as x_1, y_1 x_c, y_c and x_3, y_3. The three equations in the three unknowns are

$$y_1 = c + p\ e^{qx_1}$$
$$y_c = c + p\ e^{q(x_1 + z)}$$
$$y_3 = c + p\ e^{q(x_1 + 2z)}$$

where $z = x_c - x_1$. A solution to these equations is as follows:

$$q = \frac{1}{x_c - x_1}\ \log_e \frac{y_c - y_2}{y_1 - y_c}$$

$$p = \frac{y_1 - y_c}{e^{qx_1} - e^{qx_c}}$$

$$c = y_1 - pe^{qx_1}$$

When the least-squares refinement is started with these initial values it should converge rapidly. The *i*th row of the Jacobian is given by

$$[J_i] = [1\ e^{qx_i}\ px_i e^{qx_i}]$$

We may compare exponential fitting with the linearized procedure outlined in Section 7.6.4. To enable the linearization to proceed we need further information. When the exponential parameter is negative this may be obtained by setting x to infinity (when it is positive we set x to zero):

$$y_\infty = c + p\ e^{-\infty}; \qquad c = y_\infty$$
$$\log_e(y_i - y_\infty) = \log_e(p) + qx_i$$

We now have an expression that is linear in x.

The extra information represented by y_∞ must be obtained experimentally, and so is subject to experimental error. Since y_∞ is to be subtracted from every actual observation, the new 'observations' $\log_e(y_i - y_\infty)$ will have non-zero covariance even when the errors on y are uncorrelated. Both the variances and covariances of the transformed observations are difficult to calculate because the transformation is non-linear. Thus we can see that although the linearized form is convenient for hand-calculation it suffers from two drawbacks: it requires a measurement of y_∞ and correct weights are difficult to calculate and strictly speaking the weight matrix cannot be taken as diagonal.

In contrast to fitting of the linearized model, direct fitting of the exponential has the advantage of not requiring extensive data collection. When fitting an exponential decay adequate results can usually be obtained from data collected over about three half-lives, that is $(y - y_\infty)$ decreases to about an eighth of its original value. If the data are not noisy an even shorter period of data collection will be satisfactory.

8.2.2 MULTIPLE EXPONENTIALS

A model for a multiple exponential can be written as[*]

$$y = c + \Sigma p_j \, e^{q_j x}$$

Of all least-squares models none has given more trouble than multiple exponentials. The undamped refinement almost always diverges, and refinement using a Marquardt parameter often fails. One of the fundamental reasons why the refinement is so ill-conditioned is the failure of the linear approximation, given in Section 4.1 and adapted for the multiple exponential model as follows:

$$\Delta y^{\text{calc}} = \frac{\partial y}{\partial c} \Delta c + \Sigma \frac{\partial y}{\partial p_j} \Delta p_j + \Sigma \frac{\partial y}{\partial q_j} \Delta q_j$$

$$+ \frac{1}{2} \left(\Sigma\Sigma \frac{\partial^2 y}{\partial p_j \, \partial q_k} \Delta p_j \, \Delta q_k + \Sigma\Sigma \frac{\partial^2 y}{\partial q_j \, \partial q_k} \Delta q_j \, \Delta q_k \right) + \cdots$$

Since the derivatives with respect to the exponential parameters are given by

$$\frac{\partial y}{\partial p_j} = e^{q_j x}, \qquad \frac{\partial y}{\partial q_j} = p_j x \, e^{q_j x}$$

we can derive the expression for the second derivatives:

$$\frac{\partial^2 y}{\partial q_j \, \partial p_j} = x \, e^{q_j x}, \qquad \frac{\partial^2 y}{\partial q_j^2} = p_j x^2 \, e^{q_j x}, \qquad \frac{\partial^2 y}{\partial p_j^2} = 0$$

---

[*] Such models are not uncommon. For example, if a substance A decays by first-order kinetics to a substance B, which in turn decays by first-order kinetics to a substance C, the concentration of C, is given by the expression

$$[C] = [A_0] + \frac{[A_0] k_2}{k_1 - k_2} e^{-k_1 t} - \frac{[A_0] k_1}{k_1 - k_2} e^{-k_2 t}$$

where $[A_0]$ is the initial concentration of A, k_1 and k_2 are the rate constants for the two reactions and t is time.

Truncation of the Taylor series expansion of y^{calc} at the first order is to ignore these second derivatives, which are of the same order of magnitude as the first derivatives. This cannot ever be satisfactory, and the refinement only converges when the parameter shifts Δp_j, Δq_k are small, making the higher-order terms in the expansion small regardless of the values of the second derivatives.

Another way of looking at the problem is to examine the partial contour map shown in Figure 4.6. This illustrates the enormous effect of the non-linearity on the contours of the sum of squares. The refinement converges when these contours become approximately elliptical, and that occurs in a very small region of parameter space. There is also the problem of multiple minima.

The conclusion that may be drawn from this analysis is that fitting of multiple exponentials can only succeed in general if the initial parameter estimates are very good. No amount of algebraic juggling can alter the fact that the system is highly non-linear. It has been claimed (Krogh, 1974) that the variable projection algorithm may perform better than the standard one, but this claim is not well substantiated. So how may the parameters be estimated?

We have seen in Section 4.8.3 that the Nelder–Mead simplex method offers an excellent means of obtaining parameter estimates from arbitrary starting values. It is worthwhile to try to guess realistic starting values, but not essential. Some ball-park values can be obtained by fitting with a single exponential.

It must be stressed that high data quality is essential when fitting multiple exponentials. Some of the difficulties that people have experienced in this field can be traced back to inadequate data, usually in the form of too few data points and/or extending over a too limited range. The data should extend over about three half-lives in each of the exponential terms. Also, experimental noise must be reduced as much as possible. Even when the data are good it is still important to invest a lot of time and effort in getting good initial parameter estimates.

8.3

8.3.1 GAUSSIAN, LORENTZIAN AND RELATED FUNCTIONS

There are many types of spectroscopy which produce features that can be approximated by one of these functions. The spectrum is considered to be composed of a linear combination of one or more *bands*. We use the term decomposition to describe the process of recovering the components from the spectroscopic data. Whilst the process of decomposition by the least-squares method is relatively straightforward, the rationale for applying the process is not usually soundly based on theory. By this I mean that there is seldom a sound theory to determine the choice of the band shape function, and in many cases it is not even certain how many components should be expected of the decomposition. Thus the problems of curve decomposition are principally of a chemical or physical nature rather than of a numerical nature. Numerical problems do arise, to be sure, but their origin can usually be traced back to errors or inadequacies in the models being examined. For general reviews see Barker and Fox (1981) and Maddams (1980).

8.3.2   FITTING SINGLE BANDS

It is a commonplace observation that single bands rarely conform to the ideal Gaussian or Lorentzian shape.[*] When faced with a single band one may attempt to determine the actual shape function. This should be regarded as an exercise in empiricism, whose objective is the determination of an empirical shape function which may be of use in more complicated situations.

The simple shape functions may be written in terms of the standard variable z:

$$z = \frac{p - x}{w/2}$$

where p is the position of maximum intensity and w is the full width at half-height, FWHH. Let h be the maximum height of the band. Then

$$y = \frac{h}{1 + z^2} \quad = L \quad \text{(Lorentzian)}$$

$$y = he^{-\log_e 2\, z^2} = G \quad \text{(Gaussian)}$$

The factor $\log_e 2$ is needed in the exponent of the Gaussian in order that $y = h/2$ when $p - x = w/2$, i.e. so that the Gaussian and Lorentzian have the same half-width. The two commonest empirical functions are as follows:

$$y = fG + (1 - f)L \quad \text{(sum function)}$$
$$y = GL/h \quad \text{(product function)}$$

The sum function is to my mind preferable as it is a linear combination of Gaussian and Lorentzian, with an adjustable parameter f. Another function that has been proposed, with some theoretical justification, is the convolution of a Lorentzian with a Gaussian, which is known as the Voigt function (Sundius, 1973). A class of empirical functions that has been examined by Baker *et al.* (1978) is the class of inverse polynomials:

$$y = \frac{h}{1 + pz^2 + qz^4 + rz^6 + \cdots}$$

---

[*]Some authors prefer to use the term Cauchy shape rather than Lorentzian shape.

These functions become Lorentzians when $p = 1, q, r, \ldots = 0$, and approximate to Gaussians, when $p = 1$, $q = \frac{1}{2}$, $r = \frac{1}{6}, \ldots$.

All the functions mentioned so far are symmetrical about the point of maximum intensity, $x = p$, $z = 0$. Since there is no requirement that an empirical function be symmetrical in this fashion, various ways of introducing asymmetry have been proposed. The simplest is to allot different half-widths to the two halves of the band, $p \geqslant x$ and $p < x$. Maddams (1980) gives references to at least six other methods, but properly adds the rider that 'it is most unwise to curve fit a composite profile in terms of asymmetric peaks unless there is a rational reason for so doing'.[*]

The fitting of 'single' experimental bands is not in itself an exercise of great value. The main purpose for doing it is to get a feel for the right kind of function to use in more complex spectra.

8.3.3   MULTIPLE-BAND DECOMPOSITION

There are a number of problems associated with multiple-band spectra. There are problems resulting from extensive overlap of components, but chiefly it is often far from clear in the first place how many components are present and, if that is known, what shape functions should be used. Because of this, curve decomposition is as much of an art as a science, in which subjective judgements play an important role. Not surprisingly, many people view the results of curve decomposition with a degree of justifiable scepticism. Anyone entering the field should bear this in mind and seek to back the conclusions with independent evidence.

The general statement of the model is as follows:

$$y_i = \text{baseline}_i + \sum_j f(h_j, p_j, w_j, \ldots, x_i)$$

The baseline is considered in detail in the following section. Note that the width parameter enters the G and L expressions as a squared term, and so may give rise to multiple minima. This should be avoided by a change of variable, as described in Section 4.4.

When fitting curves composed of many bands it is essential that the least-squares refinement be commenced with good estimates of the band parameters. This means that the curve must be simulated so well that differences between the calculated curve and the simulated one should be barely visible to the eye. If necessary, the simulation can be built up, starting with two bands, performing a least-squares fit, adding a third band and so on. Unless the initial

---

[*] If the instrument transfer function is known it is valid to fit data from this instrument with a function that is a convolution of the transfer function and a simple band shape. Thus, if the instrument transfer function is exponential, observed bands will be skewed. This is common in chromatography.

band parameters are close to the ultimate values the refinement process may be very unpredictable. This is a suitable case for treatment by chaos theory which deals, amongst other things, with the sensitivity of a non-linear process to initial parameter values.

To illustrate the techniques involved in curve decomposition, let us consider the spectrum shown in Figure 8.3. This spectrum shows the absorption due to the $C-N$ stretching mode of the thiocyanate ion. The spectrum was obtained from a solution of lithium thiocyanate dissolved in dimethyl formamide at a concentration of 0.2 mol dm^{-3} at a temperature of 5°C, and is one of an extensive series of spectra with five solute concentrations and six temperatures. The two main regions of absorption correspond to the thiocyanate ion in two different environments: one a solvent environment and the other an environment in which the thiocyanate ion is in contact with a lithium ion. In fitting this spectrum we should bear in mind that it is one of a series and that whatever model we choose should be applicable to all the other members of the series. The spectrum is deceptively simple. As the temperature is raised the

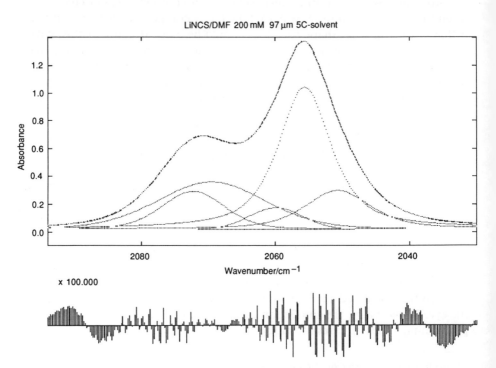

Figure 8.3. Decomposition of a spectroscopic curve into components corresponding to model 7 in Table 8.1. Upper box: data and components. Lower box: magnified residuals

band maximum at ca. 2072 cm^{-1} moves markedly to lower wavenumbers and the band maximum at ca. 2056 cm^{-1} moves slightly to higher wavenumbers. This suggests that both the main features in the spectrum are composite bands and that the shift in observed maximum is due to changes in the relative intensity of the components, which are so close together that separate peaks cannot be observed.

The spectrum has been decomposed into two, three, four and five components, and some of the results are shown in Table 8.1. For the purpose of decomposition the range fitted was 2100–2036 cm^{-1}, an interval containing 321 equally spaced data points. A Gauss–Lorentz sum function with a refinable fraction of the Gaussian component was used throughout, and the baseline was fixed at zero. It is model number 8 that is shown in the figure.

The first six models tried were unsatisfactory in one way or another. Of these model 6 was the best, but the residuals still showed distinct systematic trends. So how can we decide whether bands *b* and/or *c* are genuine or are artefacts of the decomposition process? To do this we need some independent evidence.

One worthwhile check is to compare the derivatives of the model with the derivatives obtained by numerical differentiation of the data. For model 8 the derivatives obtained by the two methods are indeed very similar to each other, as shown in Figure 8.4. A maximum difference of 5% between the two derivatives is acceptable in the light of the fact that the numerical derivative is itself only approximate.

Variation of the relative intensities of bands *c* and *d* with temperature could account for the small shifts observed in the position of maximum absorption. Likewise, variation of the relative intensities of bands *a* and *b* could account for the movement of the first maximum. Unfortunately, the position of band *b* is not well defined and the relative intensities of *a* and *b* are also only to be

Table 8.1. Selected results from the decomposition of the spectrum shown in Figure 8.3

| Model | Band position/cm^{-1} | | | | | R.m.s. residual |
	a	*b*	*c*	*d*	*e*	
1	2071.8			2055.6		5.956×10^{-3}
2	2073.5	2069.5		2055.6		5.621×10^{-3}
3	*		*	*		*c, d* bad shapes
4	*	*		*		*a, b* bad shapes
5	*			*	*	Collapsed to model 1
6	2072.6	2066.6		2055.6	2048.1	1.915×10^{-3}
7	2072.1		2060.8	2055.6	2050.3	9.183×10^{-4}
8	2071.6	2072.4	2061.3	2055.7	2050.0	9.111×10^{-4}

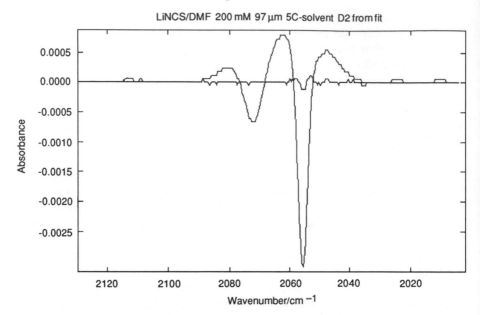

Figure 8.4. Comparison of the second derivatives obtained by convolution and curve decomposition. The data and second derivative refer to model 8 in Table 8.1. The line centred at $y = 0$ shows the difference between the derivative of the fitted curve and the derivative obtained by convolution. The steps on the curves reflect the fact that the original data have limited precision

guessed at. The reason for this must be that if there are two bands underlying the peak at ca. 2070 cm^{-1}, they must be so extensively overlapped as to make decomposition impossible.

If we are to accept that bands b and c are genuine then we must also provide some logical explanation for their existence; in other words we must develop a theory to account for the observations. There is a great danger here of making a circular argument—the theory is only meaningful if the presence of these bands has been established independently of theory.

This example shows some of the difficulties that may be encountered in the method of curve decomposition, and justifies the statements made earlier concerning the caution with which the results must be treated. The major problem is that of band overlap which is discussed in Section 8.3.5.

8.3.4 BASELINE FUNCTIONS

There are three main choices for baseline functions. One is a polynomial

$$b_i = p_0 + p_1 z_i + p_2 z_i^2 + \cdots$$

and another is the exponential

$$b_i = p_0 + p_1 \, e^{p_2 z_i}$$

The third option is to use the tail of a Gaussian or Lorentzian. Notice that the baseline functions have been written in terms of a subsidiary variable z. This should be chosen in such a way that the baseline parameter values, and more particularly the Jacobian elements, should be comparable to those of the band parameters.

8.3.5   THE BAND OVERLAP PROBLEM

If we have two overlapping bands we can define the degree of overlap as the ratio of the separation of the peak maxima to the average of the half-widths. Whether or not decomposition is possible depends both on the degree of overlap and on the amount of noise in the data. Obviously, if there were no noise in the data, and if there were no limitations of numerical precision in the computer, it should be possible to separate components with any degree of overlap. In the real world the numerical precision involved in the calculations is of minor importance.

It is also possible to describe the degree of overlap without a knowledge of the half-widths. When the degree of overlap of two bands is small the two peaks will be clear to the eye. As they are brought closer together there will come a point when the bands cease to give separate peaks, but a shoulder is visible, with a point that has zero gradient. This is also an inflection point. We may therefore define a first *critical separation* as that separation for which slope and curvature are zero at some point:

$$\frac{dy}{dx} = \frac{d^2y}{dx^2} = 0$$

In fact we may define a whole series of critical values in like fashion, and call them the second, third, etc., critical separations:

$$\frac{d^n y}{dx^n} = \frac{d^{n+1} y}{dx^{n+1}} = 0, \qquad n\text{th critical separation}$$

Vanderginste and de Galan (1975) have calculated the first and second critical values for pairs of Gaussian and pairs of Lorentzians as a function of the relative separation and half-width of the components. They showed that the critical separations increase as the ratio of the two band heights increases; i.e. it is more difficult to decompose a spectrum consisting of a major peak and a minor peak than to decompose one with two peaks of similar intensity. Vanderginste and de Galan suggested that decomposition would only be reliable if the separation of the peaks was larger than the second critical value. Whether or not this is so in an experimental spectrum can be ascertained by inspecting the second derivative of the spectrum. If the maximum between two minima corresponding to absorption maxima has a positive value the separation is larger than the second limit.

The above-mentioned authors tested their ideas on experimental data obtained with an infrared spectrophotometer whose design (it was an optical null instrument) imposed a limit on the precision of the data of $S/N = 100$.[*] We have pointed out (Gans and Gill, 1980; this paper gives graphical illustrations of the first and second critical separations) that with more precise data reliable decomposition is possible at smaller band separations. The critical separations for a pair of bands of equal height and equal half-width are given in Table 8.2. We have shown that with a noise level defined as $S/N = 500$ reliable decomposition may be possible if the band separation is larger than the fourth critical value. The figures give a rough guide to the possibility of successful curve decomposition. Bands of Gaussian shape are easier to separate because the intensity of the band tail drops off more sharply than in the Lorentzian case. The equal half-width case is the most difficult, and resolvability improves as the ratio of half-widths increases. The equal height case is the easiest, and resolvability worsens as the ratio of heights increases.

What happens when one tries to decompose bands that are really too close together? In the first place the resultant band parameters will be in error. More importantly, the estimates of error in these parameters will be too low, giving one a false estimate of the likely ranges of the parameters. Secondly, it will be found that some parameters are highly correlated. When this happens it is a good indication that the fit is not reliable. Also, if the parameter correlations are very high the least-squares refinement will become ill-conditioned, and it may be observed that an ever-increasing Marquardt parameter is required as the refinement proceeds. Lastly, because of the high correlations convergence becomes somewhat arbitrary—there will be a range over which alternative sets of parameter values give effectively equally good fits, and this range will be much larger than the standard deviations on the parameters themselves.

Table 8.2. Critical separations for two equal bands

	Lorentzians	Gaussians
Second	0.59	0.58
Fourth	0.43	0.32
Sixth	0.37	0.23

---

[*] S/N stands for signal-to-noise ratio. It may be defined as the maximum signal divided by the r.m.s. noise, that is $S/N = 100$ corresponds roughly to a 1% noise level. S/N is a crude measure of the noise level as it takes no account of the way noise varies with the signal intensity.

It is useful to monitor the progress of the refinement by plotting the calculated bands for each refinement cycle. This will indicate whether the fit is sensitive to correlated changes in the bands.

8.4 TRIGONOMETRIC FUNCTIONS (HARMONIC ANALYSIS)

A general model for harmonic analysis is as follows:

$$y_i = \text{baseline} + \sum_j \sin(p_j x_i + \Psi_j)$$

This model may effectively include cosine terms via the phase angle Ψ. Harmonics can be explicitly included in the model by writing it as

$$y_i = \text{baseline} + \Sigma\, k\, \sin(p_{jk} x_i + \Psi_j)$$

The ith row of the Jacobian for this model is

$$[\mathbf{J}_i] = [1 k x_i\, \cos(p_{jk} x_i + \Psi_j) \cdots k\, \sin(p_{jk} x_i + \Psi_j) \cdots]$$

Guest (1961) has given a detailed discussion of harmonic analysis. It is to be noted (Guest, Section 10.3) that if the data points are equally spaced the summations involved in the elements of the normal equations matrix can be written down in closed form. De Boor (1978) also has a brief discussion of harmonic analysis in terms of periodic spline functions.

The problem of multiple solutions has already been alluded to in Section 4.4. If p_j is a parameter solution, then so are $p_j + 2\pi n$, $n = 1, 2, 3, \ldots$. This is perhaps not as serious as it appears at first sight as the solutions are separated by 2π.

Sines and cosines can now be expressed in terms of complex exponentials:

$$\sin(\theta) = (e^{i\theta} + e^{-i\theta})/2i$$

so one might expect that complex harmonic analysis might be bedevilled by the same problems as beset the fitting of sums of exponentials. In any case, a Fourier transform technique is probably more apposite, as the Fourier transform of a sine wave is simply an impulse function.

8.5 FITTING TO SURFACES

If there are two independent variables, x and z, a plot of the observations y as a function of x and z is a surface in three dimensions. Such a system presents no special difficulties as far as the least-squares determination of parameters is concerned, but it does present real difficulties in respect of visualization.

With more than two independent variables the observations lie on a hyper-surface. The following example illustrates how a system with two independent variables may be handled.

A chemical reaction $X \rightarrow Y$ is known to follow first-order kinetics so that, at constant temperature, the concentration of the reactant X can be expressed as follows:

$$\left(\frac{\partial [X_{tT}]}{\partial t}\right)_T = -k[X_{tT}]$$

$$[X_{tT}] = [C_T]e^{-kt}$$

The experiments consist of measuring, at a series of times t, the concentration of X in a mixture at temperature T. $[C_T]$ is the concentration of X at time $t = 0$ for temperature T, which cannot be measured experimentally and is therefore a parameter; k is a rate constant. The experiments are performed at a number of fixed temperatures and it is expected that the rate constant varies according to the Arrhenius expression

$$k = A \ e^{-E/RT}$$

A and E are also parameters to be determined. The rate constant at any temperature can be calculated when A and E are known.

The observations are placed in a vector. The order of the observations in this vector is arbitrary, but the dependence of the observations on parameters and independent variables is noted; i.e. each index i of y_i^{obs} corresponds to certain values of t and T. Each row of the Jacobian contains the $2 + j$ partial derivatives, where j is the number of experiments:

$$\frac{\partial [X_{tT}]}{\partial A} = \frac{\partial [X_{tT}]}{\partial k} \frac{\partial k}{\partial A} = -[C_T]t \ e^{-kt - E/RT}$$

$$\frac{\partial [X_{tT}]}{\partial E} = \frac{\partial [X_{tT}]}{\partial k} \frac{\partial k}{\partial E} = \frac{[C_T]tA}{RT} e^{-kt - E/RT}$$

$$\frac{\partial [X_{tT}]}{\partial [C_T]} = e^{-kt}; \qquad T = T_1 \text{ to } T_j$$

The whole Jacobian will have the form shown opposite.

In this example it is convenient to group the data at each temperature together and within each group to order the data in the natural time sequence. Note that the partial derivatives with respect to initial concentration are zero except in the data from one temperature.

In general one may choose any ordering that happens to be convenient. The chosen ordering defines how the Jacobian is to be calculated. The normal equations can then be set up and solved following the usual procedure.

	$\dfrac{\partial[X_{tT}]}{\partial A}$	$\dfrac{\partial[X_{tT}]}{\partial B}$	$\dfrac{\partial[X_{tT}]}{\partial C_1}$	$\dfrac{\partial[X_{tT}]}{\partial C_2}$	..	$\dfrac{\partial[X_{tT}]}{\partial C_j}$	
$t = t_{11}$	x	x	x	0	..	0	$T = T_1$
.	.	.	.	.	..	.	
$t = t_{m1}$	x	x	x	0	..	0	
$t = t_{12}$	x	x	0	x	..	0	$T = T_2$
.	.	.	.	.	..	.	
$t = t_{m2}$	x	x	0	x	..	0	
.	.	.	.	.	..	.	
.	.	.	.	.	..	.	
$t = t_{1j}$	x	x	0	0	..	x	$T = T_j$
.	.	.	.	.	..	.	
$t = t_{mj}$	x	x	0	0	..	x	

Thus far fitting a surface is no different from fitting a line, except for the greater complexity of indexing. Problems arise when we want to plot the residuals, as the plot is essentially multidimensional. In the example given the data fits naturally into groups in which only one independent variable changes (time) whilst the other remains constant (temperature). One may then plot the residuals at each temperature as a function of time. If the independent variables change together there is no alternative but to tabulate the results and use purely numerical procedures to assess the quality of the fit.

An example with three independent variables occurs in structure determination by single crystal X-ray diffraction. The structure factor is given by

$$F_{hkl} = \sum_i f_i \exp[2\pi i(hx_i + ky_i + lz_i)]$$

and measured diffraction intensities are proportional to F_{hkl}^2. The independent variables are h, k and l; the parameters are x_i, y_i and z_i, the fractional Cartesian coordinates of the ith atom in the unit cell.

Surfaces can be fitted with empirical functions. The method of orthogonal polynomials (Section 7.3.1) has been extended to orthogonal multinomials by Bartels and Jeziorawski (1985). These authors have derived a three-term recurrence relation for the calculation of the multinomials and have included an algorithm in their publication.

8.6 FUNCTIONS OBTAINED EXPERIMENTALLY

A form of curve decomposition is used to analyse mixtures when the functional form of the individual components can be determined experimentally. Suppose that a chemical process gives a product that is a mixture of two chemical species, A and B, that can be separated from each other. An absorption spectrum of the mixture, consisting of absorbances as a function of wavelength, may be measured. If the spectra of the components A and B at unit concentration are known, and are denoted as $s(A)$ and $s(B)$, the spectrum of a mixture can be decomposed to yield the concentrations, c_A and c_B, of the components by specifying the model

$$y_i = c_A s(A)_i + c_B s(B)_i$$

where c_A and c_B are the parameters and i is an index referring to wavelength. Note that this model is linear in the least-squares sense and that it is a model in which there are two independent variables, $s(A)$, $s(B)$, so that it is a form of surface-fitting (Section 8.5). The ith row of the Jacobian is

$$[\mathbf{J}_i] = [s(A)_i \ \ s(B)_i]$$

The wavelengths chosen for the measurements should be such that the measured absorbances show the maximum sensitivity to changes in the parameters. In practice this means making at least two measurements, each at or near the wavelength of maximum intensity in the spectra of one component.

Two points are easily overlooked. Firstly, the independent variables are the quantities $s(A)_i$, $s(B)_i$ which are determined experimentally and are therefore subject to error. Moreover, if a number of wavelengths is used, there is often correlation between values such as $s(A)_i$ and $s(A)_k$ where $k - i$ ranges from 1, for adjacent data points, upwards. Neglect of the errors in the independent variable will lead to the errors on the parameters being underestimated, and neglect of correlation could make matters even worse. The problem can be eased experimentally by ensuring that the noise on the standard spectra $s(A)$ and $s(B)$ is reduced to a level well below the noise on the spectra of mixtures.

The other point concerns baseline errors. The model given above assumes that the quantities y_i, $s(A)_i$ and $s(B)_i$ are correct measures of the distances from the baseline or, put another way, that the baseline is everywhere zero. This is experimentally very difficult to achieve to the precision of the measurements. Neglect of baseline error introduces a systematic error into the model. A better model would be one in which the baseline heights, $b(y)$, $b(A)$ and $b(B)$, are regarded as parameters:

$$y_i = b(y) + c_A [s(A)_i + b(A)] + c_B [s(B)_i + b(B)] + \cdots$$

However, this model is now non-linear in the parameters. Also, the baseline parameters will be very small. An alternative procedure is to use the model

$$y_i = b(y) + c_A s(A)_i + c_B s(B)_i$$

but to take baseline error into account when calculating the errors on the standard spectra. All these attempts to allow for baseline error assume that the baseline is flat, i.e. it is independent of x. This is also a questionable assumption in regard to real spectrophotometers.

More complicated mixtures can be analysed in the same way as the two-component mixture by specifying a model such as

$$y_i = \sum_j c_{ij} s_{ij}$$

Since the form of the spectrum of each component is known it would be sensible to make measurements at wavelengths corresponding to the maximum absorption of each component.

With the more complicated mixtures it is not always obvious whether a particular component is present or absent. If a component is absent the least-squares calculation will probably fail. There is a technique for determining the number of components that may be present. It involves recording the spectra of a number of mixtures. Each spectrum is a linear combination of the spectra of the components. Therefore if more mixtures are examined than there are components in them the spectra of the mixtures will not be linearly independent. The determination of the number of linearly independent spectra, which is equal to the number of components, is an aspect of principal components or factor analysis.

Let us label the observations as y_{jk}, where j is an index that relates to wavelength and k is the spectrum number. The observations form a rectangular matrix:

$$\mathbf{y} = \begin{bmatrix} y_{11} & y_{12} & \cdots & y_{1n} \\ y_{21} & y_{22} & \cdots & y_{2n} \\ \cdots & \cdots & \cdots & \cdots \\ y_{m1} & y_{m2} & \cdots & y_{mn} \end{bmatrix}$$

The number of linearly independent columns in this matrix is equal to the number of its non-zero singular values[*] (see Appendix 2). More details concerning factor analysis can be found in Malinowski and Howery (1989).

A way of checking for the presence of a component is to use cross-correlation (Section 9.3). The observed data are cross-correlated with the known spectrum of a component. If the component is present in the mixture the function will be correlating with itself, whereas if the component is absent there will not be any self-correlation.

---

[*] Traditionally the problem was handled by finding the eigenvalues of $\mathbf{yy}^T$ which are the squares of the singular values of $\mathbf{y}$.

9

Fourier Transform Techniques

9.1 INTRODUCTION

Fourier transforms are now used routinely in many kinds of physical measurement. We have encountered one already in connection with n.m.r. spectroscopy (Section 5.3), and others include FTi.r. (infrared) and FTm.s. (mass spectrometry). We must therefore consider data that have been subjected to Fourier transformation, particularly in relation to noise, but also in relation to the instrument transfer function. Fourier transformation is also a valuable tool for theoretical data analysis, and as such needs to be considered in its own right.

The essence of the Fourier transform is that the data can be represented in regions of different physical dimensions. The two regions are known as the *domain* and *codomain*. We shall call the variable in the codomain the *covariable*. The dimensions of the covariable are the reciprocal of the dimensions of the variable in the domain. The two commonest examples are [time] and [length], the square brackets being taken to imply dimension. The reciprocal of [time] is [frequency]; i.e. if a variable has units [seconds], its covariable will have units [per second], or [hertz]. The reciprocal of [cm] is [wavenumber per cm], or [cm^{-1}]. The choice of domain or codomain is a matter of convention and neither is more significant than the other. For convenience it is common to place experimental measurements in the domain and transformed measurements in the codomain.

Consider a set of data collected in the time domain, and suppose that this data can be represented by a function. For example, the n.m.r. free induction decay (f.i.d.) can be considered to be a function composed of a sum of exponentially damped cosine waves. We shall ignore for the moment the fact that the experimental data consist of samples of this function and that experimental data have added noise. The functional relationship can be written as

$$y_t = f(t)$$

and can be expanded as a Fourier series of sine and cosine terms in a covariable v:

$$y = \sum_{v=0,\infty} a(v)\cos(vt) + b(v)\sin(vt)$$

The coefficients $a(v)$ and $b(v)$ are cosine and sine Fourier transforms of y. In plain words, the coefficients show how much each term $\cos(vt)$ and $\sin(vt)$ contributes to the sum which overall makes up the function. Thus, a plot of $a(v)$ as a function of v displays the intensity of each cosine contribution. Such a plot is often called the power spectrum of the function of time, since it shows the intensity, or power, in each frequency component. Performing the Fourier transform is similarly called spectrum analysis. In the context of this book such terminology would be very confusing in relation to a spectrum such as the infrared absorption spectrum, so it is not used.

The value of Fourier transformation lies in the fact that *each* data point collected in the time domain contains information on *all* the transformed data points in the frequency domain. This may be termed the *multiplex advantage* of the method.[*] In FTn.m.r. spectroscopy the f.i.d. may consequently be collected in a much shorter time than the spectrum in the frequency domain. This leads, via coaddition, to a much higher signal-to-noise ratio for the use of the same amount of time as would be needed to collect a single spectrum in the frequency domain.

The calculation of the Fourier transform is a numerical process and is usually carried out by means of the fast Fourier transform (FFT). However, it is useful to understand what is involved in the FFT process, and particularly so in relation to the effect of the process on experimental noise.

The functions $F(z)$ and $f(x)$ are said to be Fourier transforms of each other[†] when

$$F(z) = \int_{-\infty}^{\infty} f(x)e^{-ixz}\, dx = FT^{-}[f(x)]$$

$$f(x) = \frac{1}{2\pi} \int_{-\infty}^{\infty} F(z)e^{ixz}\, dz = FT^{+}[F(z)]$$

---

[*] The diffraction pattern obtained in single-crystal X-ray diffraction is an example of a Fourier transform that occurs naturally—the diffraction pattern is a transform of the electron density in the crystal. The intensity of each diffraction spot provides information on the positions of many atoms in the crystal.

[†] In this section we use the convention that upper and lower case variables are Fourier transforms of each other.

The complex Fourier transform can be rewritten by using the identity $e^{ix} = \cos(x) + i\,\sin(x)$ as

$$f(x) = \frac{1}{2\pi} \int_{-\infty}^{\infty} F(z)\,[\cos(xz) + i\,\sin(xz)]\ dz$$

$$= \frac{1}{2\pi} \int_{-\infty}^{\infty} A(z)\cos(xz)\ dz + \frac{1}{2\pi} \int_{-\infty}^{\infty} B(z)\sin(xz)\ dz$$

$$A(z) = \frac{F(z) + F(-z)}{2}$$

$$B(z) = \frac{F(z) - F(-z)}{2i}$$

$A(z)$ and $B(z)$ are known as the real and imaginary parts of the complex transform, and also as the cosine and sine transforms. A phase angle Φ may be defined as

$$\tan[\Phi(z)] = -iB(z)/A(z)$$

and using this definition we may write down a third form of the Fourier transform:

$$f(x) = \frac{1}{2\pi} \int_{-\infty}^{\infty} C(z)\cos[xz + \Phi(z)]\ dz$$

$$C(z) = A(z)/\cos[\Phi(z)] = -iB(z)/\sin[\Phi(z)]$$

When the Fourier transform is applied to experimental data the infinite integral has to be replaced by a finite summation:

$$Y(z) = \sum_{j_{min}, j_{max}} y(x_j)e^{-ix_j z}$$

The main consequence of this is that the data set, instead of ranging to infinity, has fixed limits. This is equivalent to having multiplied the theoretically infinite data set by a box function:

$$f_{box}(x) = 1, \qquad x_{min} \leqslant x \leqslant x_{max}$$

$$f_{box}(x) = 0, \qquad x < x_{min} \quad \text{and} \quad x > x_{max}$$

$$Y(z) = \sum_{j=-\infty,\infty} f_{box}(x_j)y_\infty(x_j)e^{ix_j z}$$

If the actual data set differs significantly from zero at its limits, the multiplication by the box function will cause the transform to be somewhat distorted (see Section 9.3).

9.2   ALGEBRAIC PROPERTIES OF FOURIER TRANSFORMS

9.2.1   EVEN FUNCTIONS

If $F(z)$ is an *even* function of z, that is $F(z) = F(-z)$, $B(z)$ is zero and the Fourier transform has no imaginary part; the cosine transform can be used:

$$f(x) = \frac{1}{2\pi} \int_{-\infty}^{\infty} F(z)\cos(xz)\ dz$$

$$= \frac{1}{\pi} \int_{0}^{\infty} F(z)\cos(xz)\ dz$$

9.2.2   THE LINEAR PROPERTY

The Fourier transform of a sum is the sum of the individual transforms:

$$FT[F(z) + G(z)] = FT[F(z)] + FT[G(z)] = f(x) + g(x)$$

This important property underpins the use of Fourier transforms, as it means that any signal that is composed of a linear combination of components transforms in the same way as if each component were transformed separately and the transforms combined; i.e. the transform of the sum of two Lorentzians is the same as the sum of the transforms of each Lorentzian. Also, if the data set is considered to be composed of a signal and some superimposed noise, the transform is the same as if the signal were transformed, the noise were transformed and the two were added:

$$y(t) = f(t) + \text{noise}(t)$$
$$Y(\nu) = F(\nu) + \text{NOISE}(\nu)$$

White noise may be defined as noise composed of all possible cosine waves, each wave contributing equally; i.e. the Fourier transform of white noise is a straight line with zero slope. Therefore, if the noise in the original data is white noise, the Fourier transform will merely be displaced on the Y axis. On the other hand, if the noise is not white noise the power spectrum of the noise will be added to $F(\nu)$.

9.2.3   MULTIPLICATION BY A CONSTANT

$$FT[cF(z)] = c\ FT[F(z)] = cf(x)$$

This property also has important implications. For example, in the infrared spectrum of a solution the absorbance is proportional to the solute concentration (Beer's law). It follows that in FTi.r. the intensity in the interferogram is also proportional to concentration, so the transformed FTi.r. data will obey Beer's law.

9.2.4   CHANGE OF SCALE OF VARIABLE

When the scale of a variable in the domain is changed, the covariable changes by the inverse factor:

$$FT[F(cz)] = \frac{1}{|c|} f\left(\frac{x}{c}\right)$$

One consequence of this is that if the width of a function is increased, the width of its Fourier transform is decreased.

9.2.5   SHIFTING

A linear shift in one variable introduces an exponential factor into the transform:

$$FT[F(z+s)] = e^{isx}f(x)$$

The origin of the phase correction problem in FTn.m.r. is the location of the time $t = 0$ at which the free induction decay begins. The single parameter, s, may be computed from the sine and cosine transforms.

9.2.6   DIFFERENTIATION

$$FT\left(\frac{d^n}{dz^n} F(z)\right) = (ix)^n f(x)$$

9.2.7   MULTIPLICATION

The Fourier transform of a product of two functions is the *convolution* of the individual transforms:

$$FT[F(z)G(z)] = f(x) \otimes g(x)$$

The converse of this is that the Fourier transform of a convolution of two functions is the same as the transform of the product of the two functions. Another way of putting this is to say that multiplication in one domain maps to convolution in the codomain and vice versa. This is of considerable practical importance, as many types of data are convolutions of the true data with an instrument function.

Deconvolution is the opposite of convolution. Suppose that a set of data represents the convolution of the true data $f(t)$ with an instrument function

$g(t)$. The true data may be recovered by deconvolution, as discussed in more detail in Section 9.7:

$$y(t) = f(t) \otimes g(t)$$
$$FT^+ [f(t)] = F(\nu)$$
$$FT^+ [g(t)] = G(\nu)$$
$$FT^+ [y(t)] = Y(\nu) = F(\nu)G(\nu)$$
$$F(\nu) = Y(\nu)/G(\nu)$$
$$f(t) = FT^- [F(\nu)]$$

9.3.   CONVOLUTION AND CROSS-CORRELATION

Anyone who has had difficulty in understanding a message relayed over a public address system is familiar with an effect of convolution. The confusion arises because the sound from different loudspeakers arrives at the ear at different times depending on the distance of the speakers, creating a kind of multiple echo effect. This example illustrates the essential characteristic of the convolution process. A convolution is a sum of signals with varying delay or offset. The most general expression for the convolution of two functions is as follows:

$$c(x) = f(x) \otimes g(x) = \int_{-\infty}^{\infty} f(z)g(x - z) \, dz$$

When z takes discrete values the integral is to be replaced by a summation:

$$c_x = \sum_{z=-m,m} f_z g_{x-z}$$

The variable z represents two things: the delay involved in 'hearing' $g(x - z)$ instead of $g(x)$ and a variable factor, $f(z)$, by which the delayed signal is multiplied; e.g. the sound from the different loudspeakers arrives at the ear with different delays and intensities which both depend on the distance the sound has to travel.

We can also visualize the convolution as a weighted sum in which $f(z)$ is a set of weighting factors to be applied to the function $g(x - z)$. Smoothing with a five-point quadratic function (Section 7.4.2) is an example of this:

$$c_i = (-3g_{i-2} + 12g_{i-1} + 17g_i + 12g_{i+1} - 3g_{i+2})/35$$

with the variable z taking the values from -2 to $+2$.

Convolution is both symmetrical and linear:

$$f(x) \otimes g(x) = g(x) \otimes f(x)$$
$$f(x) \otimes [g(x) + h(x)] = f(x) \otimes g(x) + f(x) \otimes h(x)$$

Experimental data may be convolutions of the true data with an instrument transfer function. Let us consider the instrument transfer function of an old-fashioned scanning spectrophotometer, and in the process we shall get a graphical illustration of how convolution works. We may assume that the spectrophotometer has a monochromator with rectangular entrance and exit slits. An hypothetical beam of monochromatic light impinges on the entrance slit and, as the grating rotates, an image of the entrance slit passes over the exit slit. Light only passes through the monochromator when the image of the entrance slit lies partially on the exit slit. The amount of light passing through is proportional to the area of overlap. This is illustrated in Figure 9.1 for nine positions of the entrance slit image.

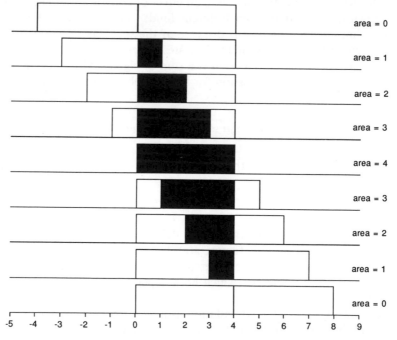

Figure 9.1. Convolution of two identical rectangle functions. The diagram can be thought of as showing the image of the entrance slit of a monochromator passing over the exit slit. The light passing through is proportional to the area of overlap (shaded). The area of overlap is proportional to the distance between the rectangles; hence the convolution produces a triangle function

When the two rectangles, representing the entrance slit image and the exit slit, do not overlap no light passes through. The amount of light passing through increases to a maximum when the two rectangles coincide, and decreases again to zero as the image passes beyond the exit slit. It is fairly easy to see that the area of overlap is an isosceles function of the displacement of the entrance slit image from the exit slit position.

To express this algebraically we denote the rectangle function corresponding to a slit of unit height and width $2a$ as

$$f(x) = 1 \quad \text{for} \ -a \leqslant x \leqslant a \quad \text{otherwise} \ f(x) = 0$$

For simplicity we may let both slits have the same function, $g(x) = f(x)$. The combined effect of the two functions is the convolution of the one with the other:

$$c(x) = \int_{-\infty}^{\infty} f(z)g(x-z) \, \mathrm{d}z = \int_{-a+x}^{a} 1 \, \mathrm{d}z = 2a - x, \qquad 0 \leqslant x \leqslant 2a$$

$$= \int_{-a}^{a+x} 1 \, \mathrm{d}z = 2a + x \qquad -2a \leqslant x \leqslant 0$$

$$= 0 \qquad\qquad x > 2a$$
$$\qquad\qquad \text{and} \ x < -2a$$

This is indeed an isosceles triangle function. In the example shown in the figure $a = 2$ so the function is $4 + x$ when x, the distance between the left-hand edges of the two rectangles, is between -4 and 0, and $4 - x$ when x is between 0 and 4.

In a real situation, the light entering the monochromator has also passed through an absorbing sample and so does not have constant intensity. Let us suppose that the sample absorption intensity is a Lorentzian function of wavelength. The output from the monochromator will be a convolution of the Lorentzian with the triangle function, because light at every wavelength is converted by the instrument transfer function into a triangle. The effect of finite slits is therefore to broaden the spectral band and to distort it from the Lorentzian shape. The degree of distortion depends on the relative half-widths of the Lorentzian and the triangle.

If the slits are narrow and diffraction effects have to be considered, it can be shown that the effect of the two slits can be approximated as a Gaussian function. A spectroscopic band that should be Lorentzian will appear as a convolution of the Gaussian slit function with the Lorentzian. This kind of function is known as the Voigt function, which has been used for fitting Raman spectra (Section 8.3.1).

Another factor that must be considered in the instrument transfer function is related to the electronic application and damping of the signal. A few years ago damping was usually done by means of a resistance–capacitance network. The equivalent instrument function for an RC network is an exponential decay. Willson and Edwards (1976) have given illustrations of the effect of this on spectra; because this function is not symmetrical about zero it causes a symmetric spectroscopic band to become asymmetric, and also causes the band maximum to appear to be shifted. In some modern instruments damping is performed by means of polynomial smoothing. With centre-point smoothing this has the advantage of not causing symmetric peak maxima to shift (see Appendix 7). It should be remembered that convolution of any signal with an instrument transfer function will cause the instrument output to have correlated errors (Section 7.4.6).

Convolution of the triangle function with the exponential gives, approximately, a skewed triangle. This is a typical instrument function in that it is obtained by successive convolution of various component functions.

In favourable circumstances the instrument function may be measured experimentally. For example, a Raman spectrometer function can be measured by replacing the sample by a laser. In such a case one might consider removing the instrument function from the data by deconvolution (Section 9.7).

Cross-correlation is formally similar to convolution:

$$c(x) = f(x) \quad \text{and} \quad g(x) = \int_{-\infty}^{\infty} f(z)g(x+z)\, \mathrm{d}z$$

It shares with convolution the property that both map to multiplication in the Fourier codomain. The applications of cross-correlation are, however, quite different. It can be used to detect a known type of signal in an unknown mixture (see Blackburn, 1970). Cross-correlation of a data set with itself produces the auto-correlation functions. For example,

$$c(1) = \sum_{i=1, n-1} y_i y_{i+1}$$

is the first auto-correlation function. If y is a random variable with mean zero the first auto-correlation coefficient gives a measure of the covariance between adjacent points.

9.4   SOME FOURIER TRANSFORM PAIRS

We may now examine some simple transform pairs. The transforms given first are the infinite transforms. Later we shall see how to derive truncated transforms from these and how to adapt these transforms to handle digitized as opposed to continuous data. The Fourier transform can be written in a number of alternative forms. The form used here is the one for which Champeney (1973) has tabulated many transform pairs. To be general, the variable in one domain is designated as x and its covariable as z.

   The first five functions are real functions whose Fourier transforms are also real. As we have seen from Section 9.2 this means that both the functions and their transforms must be *even*, i.e. symmetrical about zero. The pairs of functions given are the Fourier transforms of each other. The sixth function is a constant when x is positive, and zero when x is negative. Illustrations of all the functions given here, and many others, can be seen in Champeney.

(a) Impulse and cosine functions:

$$f(x) = \pi A \{\delta(x - x_0) + \delta(x + x_0)\}$$
$$F(z) = A \cos(x_0 z)$$

   The plots of these functions, given in Figure 9.2, are for $A = 1$ and $x_0 = \pi$.
(b) Rectangular (box) and sinc functions:

$$f(x) = A \text{ for } -x_0 \leqslant x \leqslant x_0, \qquad \text{otherwise } f(x) = 0$$

$$F(z) = 2A \frac{\sin(x_0 z)}{x_0 z}$$

   The plots of these functions, given in Figure 9.3, are also for $A = 1$ and $x_0 = \pi$.
(c) Triangle and sinc^2 functions:

$$f(x) = A(1 - |x|/x_0) \text{ for } |x| \leqslant x_0, \qquad \text{otherwise } f(x) = 0$$

$$F(z) = A x_0 \left(\frac{\sin(x_0 z/2)}{x_0 z/2} \right)^2$$

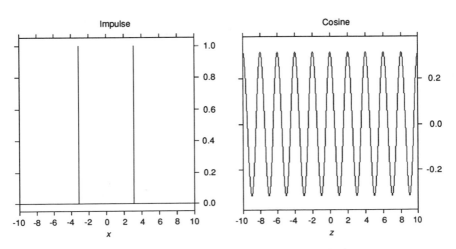

Figure 9.2. The Fourier transform pair impulse/cos

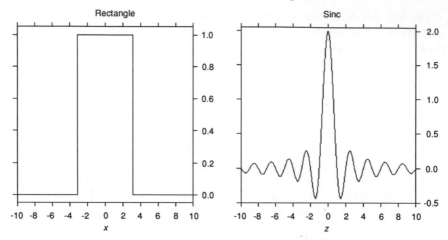

Figure 9.3. The Fourier transform pair rectangle/sinc

The plots of these functions, given in Figure 9.4, are to the same scale as the plots of rectangle and sinc functions. Note that the triangle function transforms to a broader function than does the rectangle of the same width. This has the consequence in FTi.r. that apodized spectra (q.v.) have lower resolution than unapodized spectra.

(d) Exponential and Lorentzian functions:

$$f(x) = \frac{2}{w} \frac{A}{1 + (2x/w)^2}$$

$$F(z) = e^{-w/2|z|}$$

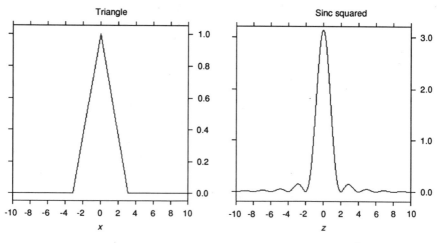

Figure 9.4. The Fourier transform pair triangle/sinc^2

The plots of these functions, given in Figure 9.5, are for $A = 1$ and $w = 1$. Note that the half-life of the exponential is inversely proportional to the half-width of the Lorentzian.

(e) Gaussian functions:

$$f(x) = A\, e^{-(2x/w)^2}$$

$$F(z) = \frac{A w \sqrt{\pi}}{2}\, e^{-(zw)^2}$$

The plots of these functions, given in Figure 9.6, are for $A = 1$ and $w = 2$. Note that a Gaussian transforms to another Gaussian, with a half-width inversely proportional to the original half-width.

(f) Heaviside function:

$$f(x) = A \qquad \text{for } x \geqslant 0, \qquad \text{otherwise } f(x) = 0$$

$$F(z) = A\pi\,\delta(0) - iA\pi/z$$

These functions are shown in Figure 9.7.

Many Fourier transforms of experimental interest can be built up from these six transform pairs by multiplication and/or convolution. Thus, multiplication of any of the first five functions by the sixth converts them to functions that are non-zero when x (or z) is positive. The FT of the product is the convolution of the FTs, and as the real part of the FT of the Heaviside function is an impulse function at zero, the real part of the FT of the product is the same as the FT of the symmetric function. To truncate a function we multiply by the rectangle, which maps to convolution by a *sinc* function in the

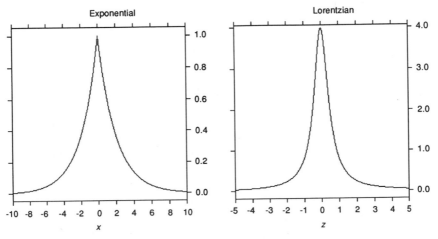

Figure 9.5. The Fourier transform pair exponential/Lorentzian

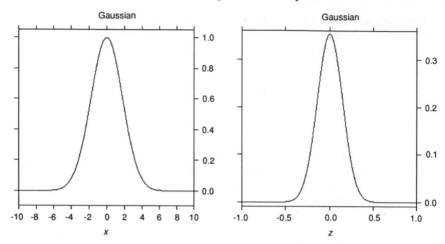

Figure 9.6. The Fourier transform pair of Gaussians

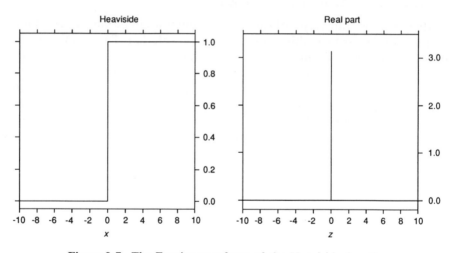

Figure 9.7. The Fourier transform of the Heaviside function

codomain. To shift a symmetric function we convolute with the impulse function (a). Thus, a Lorentzian not centred at zero transforms to a product of the exponential and cosine function, i.e. an exponentially damped cosine wave.

Some experimentally useful conclusions can be drawn from these simple transforms. The relationship between (b) and (c) confirms to us that the convolution of two rectangular functions is a triangle function. Functions (b) and (c) are very important in FTi.r. The interferogram is truncated at finite maximum retardation, equivalent to multiplying it by a rectangle function.

Consequently, the Fourier transform is a convolution of the spectrum with the sinc function. This sinc function thus becomes a component of the instrument transfer function. If the interferogram is instead multiplied by the triangle function, the spectrum is convoluted with $sinc^2$. This was called *apodization* since the $sinc^2$ function has smaller 'feet'. However, the sinc function has zeros at $z = 1/x_0$, while the $sinc^2$ function has zeros at $z = 2/x_0$, showing that it is a broader function. In other words, multiplying the interferogram with the triangle function has the (cosmetic?) advantage of reducing the false features at the feet of the spectroscopic bands, but has the disadvantage of broadening the bands. Other apodization functions have been devised.

We may deduce that the convolution of one Lorentzian with another one is also Lorentzian. The FT of a Lorentzian is an exponential, so the FT of a convolution of two Lorentzians is a product of exponentials, which is itself an exponential. Fourier transformation of the latter gives a third Lorentzian. This fact is used in a form of data processing known as resolution enhancement, which is a form of deconvolution. Dividing the FT of a Lorentzian by an exponential whose FT is a Lorentzian of smaller half-width, and back-transforming, yields a Lorentzian of greatly reduced half-width. In a spectrum composed of a number of overlapping Lorentzians application of this technique will reveal the individual components as sharp bands. Unfortunately the technique suffers from two major disadvantages. On the one hand it only works properly on Lorentzians and on the other it suffers from increased noise, as do all deconvolution techniques. Nevertheless, it is another possible tool to be used alongside differentiation (Section 7.4.4) and curve decomposition (Section 8.3) in the analysis of spectroscopic curves.

9.5 SMOOTHING REVISITED

The method of smoothing described in Section 7.4.3, based on fitting a low-degree polynomial to a subset of the data, is operationally a convolution method. We may look again at this method by examining the Fourier codomain. In the codomain the FT of the data is multiplied by the FT of the convolution function. We will therefore find the effect of smoothing on the frequencies present in the time-domain data.

Before obtaining the results we must take into consideration the effects of finite sampling. To simplify the discussion let us suppose that the data have been obtained at equal intervals of time, Δt. The actual time interval between data points determines the maximum frequency that can be observed in the Fourier expansion. This is the so-called Nyquist frequency, f_N, which is equal to $1/2\Delta t$. Any frequencies in the data which are higher than this are folded back into the expansion and appear as frequencies lower than f_N. These lower frequencies are known as *alias* frequencies.[*] The alias frequency is given by $f - m/2\Delta t$, where f is the original frequency and m is an integer. We must therefore restrict our attention to those frequencies that are less than the Nyquist frequency.

The Fourier transform of the convolution function with $2n + 1$ coefficients c_j is an example of a discrete Fourier transform, in which the integral is replaced by a summation. Moreover, since the convolution coefficient for smoothing are symmetric about zero we may use the cosine transform. Normalizing the transform to one at $z = 0$ (the sum of smoothing coefficients is always unity) it is

$$FT(z) = \sum_{j=-n,n} c_j \cos(jz)$$

$$= c_0 + 2c_1 \cos(z) + 2c_2 \cos(2z) + \cdots + 2c_n \cos(nz)$$

This shows that the transform is simply the sum of harmonics in the angle z. In order to show only that part of the transform that does not repeat itself, z is restricted to the range $0 - \pi/2$. The transform of a 15-point quadratic/ cubic smoothing function is shown in Figure 9.8, together with the transform of a Lorentzian of half-width equivalent to 20 data points. When the FT of the data is multiplied by the FT of the convolution function the effect is to attenuate the high-frequency components. In other words, the convolution selectively reduces the high-frequency components of the data. When most of them are noise, the data are not distorted to a marked extent. In the example shown the Lorentzian is hardly affected above $z = 0.1$. However, noise at or above this frequency is considerably reduced. Some distortion is, however, inevitable. Both Gaussian and Lorentzian bands transform to functions that tend asymptotically to zero, and therefore have some high-frequency components. Multiplication of these in the Fourier codomain will cause them to be attenuated, and so the data in the time domain will be distorted. The degree of distortion depends on the width of the convolution function relative to the width of the data features.

The convolution function can be described as a low-pass filter, as it passes low-frequency components of the data. The form of the filter in the low-frequency region can be seen more clearly in Figure 9.9, which shows the low-frequency part of Figure 9.8. Up to ca. $z = 0.03$, the filter is almost constant and equal to unity. This means that frequencies in this range are passed almost unchanged. The distortion becomes more noticeable at higher frequencies. In practical convolutions this would result in shaving off the top of a peak.

---

[*]The most familiar alias frequency is that of a rotating wheel when viewed on cine film. The position of the wheel is sampled 24 times a second. Because of this, if the rotation frequency of the wheel is greater than the Nyquist frequency, the wheel may appear to be rotating backwards, or may even appear to be stationary. This example is discussed in more detail in Blackburn (1970).

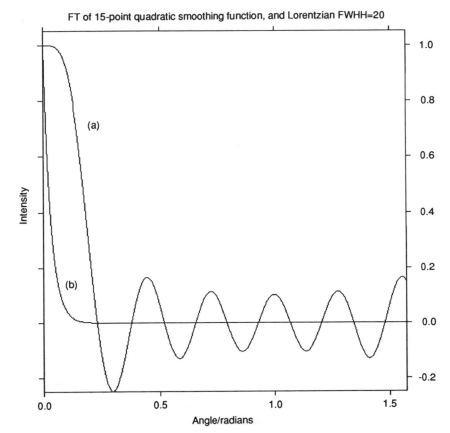

Figure 9.8. Fourier transform of (a) the 15-point quadratic/cubic smoothing function and (b) a noise-free Lorentzian of half-width equivalent to 20 points

The other important characteristic of convolution smoothing is that the low-frequency components of the data are not altered very much by the convolution process. This is fine as far as the 'true' data are concerned, but it also means that low-frequency noise is not removed. Therefore we can understand that the noise remaining after convolution (as much as half the original noise in the case of a nine-point quadratic smooth) consists predominantly of the low-frequency components of the original noise. In some cases one can actually see this in the form of ripples, particularly on derivatives.

There are lessons here regarding data collection. The broader the data interval the lower the Nyquist frequency, and hence the more inefficient will be any smoothing or differentiation by convolution. Also, the data density determines the maximum width of convolution function that can be used.

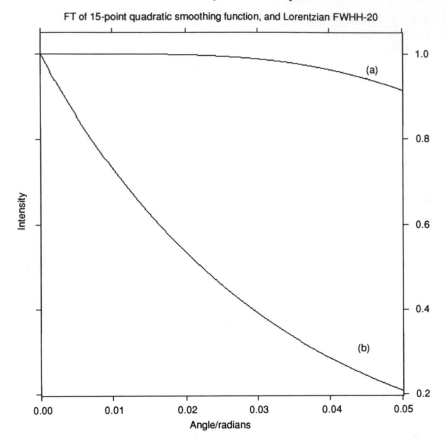

Figure 9.9. Enlargement of Figure 9.8 to show the low-frequency part

Both of these factors point towards the value of collecting as much data as possible over the region of interest.

9.6 NUMERICAL FOURIER TRANSFORMS

It is not only functions that can be subjected to Fourier transformation. Numerical data can also be transformed. The discrete transformation is a summation:

$$Y(z) = \sum_{z=-A,B} y(x)e^{-ixz}$$

A and B are the minimum and maximum values of the covariable z. The transformation is usually effected by means of a computer program that uses the fast Fourier transform (FFT) that is available in a number of places, e.g. *Numerical Recipes* (Press *et al.*, 1986).

The use of finite limits A and B may introduce undesirable complications. We have seen that if the limits are symmetrical about zero this is equivalent to multiplying $y(x)$ by a rectangle function, so that the FT is convoluted with a sinc function. If the limits are not symmetrical the transform will have an imaginary component. There will also be an imaginary component if the transform is restricted to positive values of x.

These complications have usually been taken care of in Fourier transform instruments. They become a matter of concern when data come from other sources, i.e. when the data have not been collected with Fourier transformation in mind. A typical example would be an infrared spectrum collected with a scanning spectrophotometer. In these cases the transformation of the raw data may well be unsatisfactory. For example, in transforming a Lorentzian absorption band it is necessary to extend the data with artificial tails extending to many half-widths to avoid distortion in the transform. Willson and Edwards (1976) give an illustration of this distortion, which they describe as a form of aliasing.

In general it will be necessary to adjust the data to make them suitable for Fourier transformation. The adjustments are bound to be arbitrary to some extent, and so the process cannot be recommended. In particular, truncation of the FT can be used to remove high-frequency noise and so appears to offer an alternative method of smoothing noisy data. Such a procedure is difficult to apply when the data need to be adjusted first.

9.7 DECONVOLUTION

Deconvolution may be carried out either in the domain or in the codomain. Although the two processes are equivalent they have a completely different appearance. Do not be misled by the superficial differences into believing that one is preferable to the other! A critical comparison of the various techniques used for deconvolution has been given by Cooper (1977). If a set of data is regarded as the convolution of two functions, then

$$y(x) = a(x) \otimes b(x)$$
$$a(x) = \text{FT}\,[Y(x)/B(x)]$$

To perform deconvolution we must assume that $b(x)$ or $B(x)$ is 'known'. Then $Y(z)$ is the sum of signal and noise so both components are divided by $B(z)$. The inevitable result is an increase in noise. To see this, consider two typical deconvolution functions, $b(x)$, the Lorentzian and the triangle functions, whose Fourier transforms are the decaying exponential, and sinc^2 functions respectively, both of which decay to zero at large z and whose inverse goes to infinity at large z. Multiplying by the latter increases the high-frequency noise components in the product $Y(z)$ $(1/B(z))$.

In the data domain the deconvolution is usually carried out using a modification of the Van Cittert algorithm. Blass and Halsey (1981) have discussed this in great detail. The modified Van Cittert algorithm is an iterative one, where the iteration can be written as follows, where k is an iteration number:

$$a(x)^{(k)} = a(x)^{(k-1)} + \alpha^{(k)}[y(x) - a(x)^{(k-1)} \otimes b(x)]$$

The idea is that as $a(x)$ approaches its correct form so $a(x) \otimes b(x)$ approaches $y(x)$. The iterations may be started with $a(x)^{(1)} = y(x)$. In this case we have

$$a(x)^{(2)} = y(x) + \alpha[y(x) - y(x) \otimes b(x)]$$

In the process of convoluting $y(x)$ with $b(x)$ the noise component will be magnified since $b(x)$ is of the same magnitude as $y(x)$. For this reason either α must be small or smoothing must be used. In any case, the noise continues to be magnified with each iteration cycle.

The main purpose of deconvolution is to try to remove from data the effects of an instrument function. The requirements for successful deconvolution are, however, very stringent. The instrument function must be known. The data must have a high signal-to-noise ratio. The deconvolution algorithm must include protection against a massive increase in noise. Because of these difficulties most people prefer to minimize the effects of the instrument function experimentally.

It is possible to fit data with a function which is a convolution of a Lorentzian with a triangle function (Bower, Jackson and Maddams, 1990); the convolution and its derivatives with respect to the band parameters can be obtained algebraically. The approach of fitting data with a convolution of the instrument function and the normal fitting function is preferable to deconvolution of the data since the fitting function is free from noise, and so no noise magnification occurs.

10

Potentiometric Titrations

10.1 HISTORICAL DEVELOPMENT

In this chapter we shall examine in detail one general class of calculations for complicated systems that illustrates many of the topics covered in this book. The story is also one of people as I want to give a concrete example of the way in which scientific activity is both rational and creative. Scientific development does not proceed steadily, but by fits and starts, with many a wrong turning from the road leading to the final destination.

We shall use a subclass of calculations for illustration purposes and later describe the more general approach. The subclass concerns the equilibria in solution between a metal ion, such as Ni^{2+}, a polybasic ligand, such as glycine, and hydrogen ions, H^+; the concentrations of these species will be denoted as [M], [L] and [H], with the ionic charges omitted for the sake of generality. The three reactants form complexes of various stoichiometries. The concentration of each complex is defined in terms of an equilibrium constant:

$$[M_pL_qH_r] = \beta_{pqr}[M]^p[L]^q[H]^r$$

The concentrations of the species in equilibrium are constrained by the law of mass action. This is expressed in terms of three mass-balance equations:

$$T_M = [M] + \Sigma\, p\beta_{pqr}[M]^p[L]^q[H]^r \qquad (= [M] + \Sigma\, p\,[M_pL_qH_r])$$

$$T_L = [L] + \Sigma\, q\beta_{pqr}[M]^p[L]^q[H]^r \qquad (= [L] + \Sigma\, q\,[M_pL_qH_r])$$

$$T_H = [H] + \Sigma\, r\beta_{pqr}[M]^p[L]^q[H]^r \qquad (= [H] + \Sigma\, r\,[M_pL_qH_r])$$

T_M, T_L and T_H are the concentrations in total of the metal, ligand and proton.

The objective of the calculation in the first instance is the determination of the equilibrium constants; in many cases these calculations are only part of the real problem, which is to determine the correct chemical model of the system.

The experimental procedure is as follows. A solution is prepared containing known amounts of the reactants, and to this are added portions of a solution of alkali from a burette. After each addition the hydrogen ion concentration is measured by means of a glass electrode. The experimental data therefore consists of pairs of values for the volume of alkali added, v_i, and the glass electrode measurement, E_i. We may assume that the glass electrode response is given by the modified Nernst equation:

$$E_i = E^0 + \text{slope} \times \log_e [H_i]; \qquad \text{Slope} = RT/nF \times \text{factor}$$

where E^0 is the standard potential of the electrode, obtained by calibration, and the ideal slope factor is unity. It is immediately obvious that the observations are an implicit function of the independent variable v, the volume of alkali added, as the latter does not appear in the Nernst equation. The electrode potential depends on the hydrogen ion concentration. We must therefore relate the hydrogen ion concentration to the volume of alkali added.

The hydrogen ion concentration depends on the initial masses of metal, ligand and hydrogen ions, on the initial volume, on the concentration of alkali in the burette, B_H, on the volume of alkali added and on the equilibrium constants. These dependencies are expressed by a number of equations. To simplify the notation we shall omit the index i of the titration point. The total concentrations of each reactant are given by

$$T_M = [\text{total}]_M = (\text{initial mass})_M / (\text{initial volume} + v)$$

$$T_L = [\text{total}]_L = (\text{initial mass})_L / (\text{initial volume} + v)$$

$$T_H = [\text{total}]_H = [(\text{initial mass})_H + B_H \times v] / (\text{initial volume} + v)$$

If the equilibrium constants are known, [H] along with [M] and [L] can be found by solving the three equations of mass balance.

The least-squares problem can now be stated. The observations consist of pairs of measurements, volume and electrode potentials, relevant to a set of initial masses. The parameters are the equilibrium constants. At each titration point [M], [L] and [H] are unknown quantities and it is known that in the final answer they must obey the conditions of mass balance. How does one solve such a complex system?

Historically, the first person to attempt to write a computer program for this type of problem was Lars Gunnar Sillén. Sillén's objective was to extend the study of equilibrium constants beyond what was possible using hand-calculations, and in this he was a true pioneer. With Ingri and other coworkers he developed (Dyrssen, Ingri and Sillén, 1961) the program LETAGROP, which minimized the sum of squared residuals in E. The method used became known as pit-searching (Swedish LETA (pit) GROP (search)). Essentially it is an application of Newton's method (Section 4.7.1) in which the objective function, S, is considered to be a quadratic multinomial in the parameters. If a sufficient number of S values are known the coefficients of the multinomial can be calculated, and hence the position of the minimum can be predicted. In other words, LETAGROP uses Newton's method with numerical estimates for the derivatives. The program was later modified to cope with highly correlated parameters (LETAGROP VRID; Ingri and Sillén, 1964). It was assumed for the purpose of this program that the titre volumes are free from experimental error and that measured electrode potentials had equal errors.

LETAGROP was a fantastic program, but it did not travel far from Sillén's native Sweden. In Leeds, Leslie Pettit and I considered using it in the late 1960s but were unable to do so as the program was too large for our University mainframe computer at that time. We tried to adapt it, but gave up for two reasons: we were relatively inexperienced programmers and the variable names, being in Swedish, were all but incomprehensible. The fact that the program was written in ALGOL was not a problem for us, as this was our computing department's favourite language!

Douglas Perrin, in Australia, may have had a similar experience. In any case he recognized the need for a general program to calculate stability constants. His starting point was a program called GAUSS (Tobias and Yasuda, 1963), which, as the name implies, used the Gauss–Newton method of minimization. GAUSS utilized numerical estimates for the elements of the Jacobian. The evolution progressed in stages through a three-reactant program (Perrin and Sayce, 1967) to a more general program called SCOGS (Sayce, 1968, 1971; Sayce and Sharma, 1972).

SCOGS was based on minimizing a sum of squared residuals in titre volume, v, and used as parameters log β. The philosophy underlying this program is that titre volume is considered to be the dependent variable, subject to experimental error, and the electrode potential is the independent variable, not subject to error. The hydrogen ion concentration was derived from the measured electrode potential, via the Nernst equation. The minimization was performed using the Gauss–Newton method, but again the required derivatives had to be estimated by a numerical technique. SCOGS was written in FORTRAN and was taken up by many workers throughout the world. It used a crude method of shift-cutting to guard against divergence, but sometimes failed to converge.

The next major development occurred in Italy.[*] Alberto Vacca and Antonio Sabatini (1972) published the program LEAST. Alberto had studied with Sillén and was familiar with both LETAGROP and SCOGS, but was dissatisfied with them. Together they devised a method of calculation that used the Gauss–Newton method. In this minimization the 'observations' were

---

[*] Other contributions will not be discussed here even though some were very important. A full bibliography may be found in Leggett (1985), including a full listing of SCOGS2 (Perrin and Stunzi, 1985).

taken as the derived quantities T_M, T_L and T_H, which, as we have seen, are simple functions of the added volume. The observations were given equal weight, and the hydrogen ion concentration was derived from the measured electrode potentials.

The advantage of minimizing on T_M, T_L and T_H is that the partial derivatives needed for the calculation can all be obtained from the mass-balance equations. Here are some typical examples:

$$[M] \frac{\partial T_M}{\partial [M]} = [M] + \Sigma p^2 \beta_{pqr}[M]^p[L]^q[H]^r \qquad (= [M] + \Sigma p^2 [M_pL_qH_r])$$

$$[L] \frac{\partial T_M}{\partial [L]} = \Sigma pq\beta_{pqr}[M]^p[L]^q[H]^r$$

$$\beta_{pqr} \frac{\partial T_M}{\partial \beta_{pqr}} = \Sigma p\beta_{pqr}[M]^p[L]^q[H]^r$$

There are two things to notice about these analytical formulae. The quantities on the right-hand side all reduce to multiples of concentrations of chemical species, which is very convenient for the calculation. Also, $[M]$ and $[L]$ are being treated as unknowns. A consequence of this is that the normal equations have a particular structure:

$$
\begin{bmatrix}
x & x & & & & & & & & x & . & . & x \\
x & x & & & & & & & & x & . & . & x \\
& & x & x & & & & & & x & . & . & x \\
& & x & x & & & & & & x & . & . & x \\
& & & & x & x & & & & x & . & . & x \\
& & & & x & x & & & & x & . & . & x \\
\vdots & & & & & & & & & & . & . & \\
& & & & & & & & & & . & . & \\
x & x & x & x & x & x & . & . & . & x & . & . & x \\
. & . & . & . & . & . & . & . & . & x & . & . & x \\
x & x & x & x & x & x & . & . & . & x & . & . & x
\end{bmatrix}
\begin{bmatrix}
\Delta[M]_1/[M]_1 \\
\Delta[L]_1/[L]_1 \\
\Delta[M]_2/[M]_2 \\
\Delta[L]_2/[L]_2 \\
\Delta[M]_3/[M]_3 \\
\Delta[L]_3/[L]_3 \\
\vdots \\
\Delta\beta_1/\beta \\
\cdots \\
\Delta\beta_n/\beta_n
\end{bmatrix}
= \mathbf{J}^T \Delta T
$$

In this representation zeros are shown as blank, and non-zero values as x or $..$ The normal equations were solved by using an ingenious factorization.

At this point in the story (1972) I went to Florence for a three-month sabbatical period. I had previously spent a postdoctoral year there working on far-infrared spectra with Antonio. This experience had fired my enthusiasm for computation, and on arrival I suggested that we try to calculate equilibrium constants using the Davidon–Fletcher–Powell method of minimization (Section 4.7.2). A program was accordingly written (Gans and Vacca, 1974) based on residuals in T_M, T_L and T_H, with log β as parameters, and was compared with versions of LETAGROP, SCOGS and LEAST. It was no faster nor more robust than any of them, and the use of log β as parameters was clearly seen to be less satisfactory than the use of the β values as parameters. The last point was the most significant thing that came out of that work; since the equilibrium constants appear as such in the mass-balance equations, the change of variable from β to log β makes the system *more non-linear*, and so less well-behaved in terms of refinement.

One month remained of my sabbatical. We decided to produce a completely general program, and called it MINIQUAD (Sabatini, Vacca and Gans, 1974). It was based on the methodology used in LEAST, but adapted for 2, 3, 4 or 5 mass-balance equations. We also introduced the possibility of a second indicator electrode in the hope that simultaneous measurements with a glass electrode and another ion-selective electrode would open up new experimental opportunities. The refinement was to be protected against divergence by an optimized shift-cutting technique (Section 4.5.2). We divided the programming tasks between the three of us, and using an IBM 1130 computer which we operated ourselves, by the end of the month had a program that just worked. I took the text of the program back to Leeds on punched cards. The program was then tested and debugged over the next six months.

MINIQUAD was distributed on eight-hole punched paper tape and was used by research groups in many places. Nevertheless, we were not satisfied and work continued in both Leeds and Florence. I returned to Florence again in 1974, full of enthusiasm for the improvements I had made, only to find that Alberto and Antonio had also made improvements.

I had recognized that the structure of the normal equations matrix shown above lent itself to Choleski factorization and hence to the use of a Marquardt parameter in place of shift-optimization. When a sparse matrix is factorized by the Choleski method the factor has zeros in each place where the sparse matrix has zeros. Therefore the non-zero elements of both the normal equations matrix and the Choleski factor can be stored and used in a compressed form. The original normal equations matrix has $2m + n$ rows and columns, but only $2^2m + n^2$ non-zero elements. Using the Marquardt parameter to guard against divergence was more efficient than using shift-cutting.

In Florence they 'eliminated' [M] and [L] by doing a least-squares fit to the three mass-balance equations at each titration point, taking [H] from the electrode reading. In this way [M] and [L] are no longer parameters of the least-squares minimization, but are always kept at the 'best' values for a given set of equilibrium constants. We ran the two improvements against each other but neither was a clear winner. Therefore a hybrid was constructed, MINIQUAD75 (Gans, Sabatini and Vacca, 1976), which made use of the best characteristics of both and which ran some two or three times faster than the original MINIQUAD, and was more robust.

Still we were not satisfied. A fundamental criticism was that we were not minimizing on the experimental observations, and therefore it was all but impossible to apply statistical tests to the results. The problem standing in the way of this was the fact that the observed electrode potentials are implicit functions of the parameters. To do the job properly we need the derivatives $\partial E / \partial \beta$.

It was Istvan Nagypal in Hungary who showed, from first principles, how to do the differentiation (Nagypal, Paka and Zekany, 1978). Antonio has also told me that he had worked it out independently. With the wisdom of hindsight we can see that the required mathematics is in many textbooks of advanced calculus, though it may not be in the more elementary texts (Appendix 4).

The method of differentiation is as follows. First we use the chain rule:

$$\frac{\partial E}{\partial \beta} = \frac{\partial E}{\partial [H]} \frac{\partial [H]}{\partial \beta} = \frac{RT}{nF[H]} \frac{\partial [H]}{\partial \beta}$$

Then we must solve a system of linear equations:

$$
\begin{bmatrix}
\dfrac{\partial T_M}{\partial [M]} & \dfrac{\partial T_M}{\partial [L]} & \dfrac{\partial T_M}{\partial [H]} \\[2ex]
\dfrac{\partial T_L}{\partial [M]} & \dfrac{\partial T_L}{\partial [L]} & \dfrac{\partial T_L}{\partial [H]} \\[2ex]
\dfrac{\partial T_H}{\partial [M]} & \dfrac{\partial T_H}{\partial [L]} & \dfrac{\partial T_H}{\partial [H]}
\end{bmatrix}
\begin{bmatrix}
\dfrac{\partial [M]}{\partial \beta} \\[2ex]
\dfrac{\partial [L]}{\partial \beta} \\[2ex]
\dfrac{\partial [H]}{\partial \beta}
\end{bmatrix}
=
\begin{bmatrix}
\dfrac{\partial T_M}{\partial \beta} \\[2ex]
\dfrac{\partial T_L}{\partial \beta} \\[2ex]
\dfrac{\partial T_H}{\partial \beta}
\end{bmatrix}
$$

to find the derivatives $\partial [H] / \partial \beta$. How different things might have been had Sillén used this approach!

Alberto and Antonio, aware of the advantages of minimizing directly on electrode potentials, developed a program, MIQUV, in which the derivatives were calculated by a numerical technique (see Leggett, 1985, for a listing of MIQUV). When I arrived again in Florence in 1982 they already had a draft of a program that used analytic derivatives. We then had a lively discussion on the relative merits of this new program and whether it should be developed to SUPERcede miniQUAD—and so SUPERQUAD was conceived. The fully developed program was delivered some months later (Gans, Sabatini and Vacca, 1985).

10.2 SUPERQUAD—A GENERAL PROGRAM FOR EQUILIBRIUM CONSTANT COMPUTATION

Most of the mathematics of this program have been discussed in the previous section. Here we discuss some of the details important for the full implementation of the general program. The core of the algorithm is a

Gauss–Newton minimization of a sum of weighted squared residuals in electrode potential. The refinement is protected against divergence by a Marquardt parameter (Section 4.5.3). The parameters are the equilibrium constants and other quantities that we term 'dangerous' parameters, such as the initial mass of a reactant, the concentration of reagent in the burette and the standard potential E^0. Since there are equations relating all these quantities, the required derivatives are obtained directly, or by applying the chain rule:

$$\frac{\partial E}{\partial X} = \frac{\partial E}{\partial [H]} \frac{\partial [H]}{\partial X} = \frac{RT}{nF[H]} \frac{\partial [H]}{\partial X}$$

The unknown quantities [M], [L] and [H] are to be found by solving the three non-linear mass-balance equations in these three unknowns. The refinement can therefore to be said to be subject to the non-linear constraints (Section 5.2) that the mass-balance equations are always satisfied. Although this is necessary at convergence, it is otherwise an arbitrary but useful constraint, since it removes $3m$ parameters from the least-squares refinement. Let us look at various aspects of the program in turn.

10.2.1   DATA INPUT

The data usually consists of pairs v_i, E_i. The potential may be in millivolts or pH, with a flag to indicate which choice has been made. Internally the program uses millivolts in order to apply the Nernst equation easily. Each data pair carries a flag to include or exclude it from the calculation. Parameter estimates are also needed, and this can be far from straightforward with the more complicated models. In some cases the estimates need to be very precise. Each parameter carries a refinement key according to which it is to be refined, held constant, ignored or constrained to be equal to another parameter. The constraints are checked for consistency.

The number of mass-balance equations may be 2, 3 or 4, and the number of electrodes may be 1 or 2 (when there are two electrodes the data are in the form of triplets, v_i, E_i, F_i, where E and F are the two electrode potentials). There may be more than one titration curve, and titration curves with a different number of mass-balance equations may be present together, e.g. curves from the titration of the ligand with and without the presence of metal ions. This degree of flexibility calls for more involved coding than the simple case, but it does not complicate the central minimization algorithm.

10.2.2   CALCULATION OF WEIGHTS

Both E_i and v_i are numbers obtained experimentally, and both are subject to error. We use the methods of Section 3.4.3 to construct a weight matrix.

Assuming that there is no correlation, the calculation of weights requires estimates of the errors on E_i and v_i, along with an estimate of the slope of the curve, dE/dv. It is reasonable for this type of experiment to set the errors as constants and the covariances as zero:

$$\sigma_i^2 = \sigma_E^2 + (dE/dv)_i^2 \sigma_v^2; \qquad W_{ii} = 1/\sigma_i^2$$

Notice that although σ_E and σ_v are constant the weights are not constant. This is important for the following reason. In the region of a so-called end-point the electrode potential may change very rapidly, and a small error in v could have a large effect on the measurement. This is illustrated by the complex titration curve shown in Figure 10.1. Data points near an end-point, e.g. at around

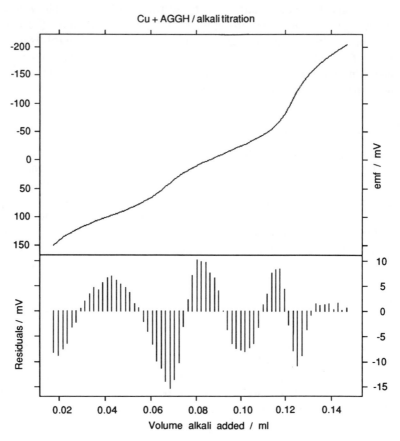

Figure 10.1. Upper box: curve of measured electrode potentials obtained from a solution containing Cu^{2+} and the tetrapeptide ala-gly-gly-his, titrated with alkali. Two end-points are clearly visible at ca. 0.7 and 0.13 ml additions. Lower box: weighted residuals given by a SUPERQUAD fit

0.125 ml, are given reduced weight by the formula. We estimate the slope by piecewise fitting of a cubic polynomial to the observed data. The data points do not have to be equidistant from each other, so the general polynomial fitting process (Section 7.1) is used.

With two electrodes the weight matrix becomes 2×2 block-diagonal, corresponding to case (c) in Section 3.4.2. There are two slopes to calculate and two variances at each point. Denoting the two potentials as E and F, each 2×2 block has the following form:

$$W_{ii} = \begin{bmatrix} \sigma_E^2 + (\partial E/\partial v)_i^2 \sigma_v^2 & (\partial E/\partial v)_i (\partial F/\partial v)_i \sigma_v^2 \\ (\partial E/\partial v)_i (\partial F/\partial v)_i \sigma_v^2 & \sigma_F^2 + (\partial F/\partial v)_i^2 \sigma_v^2 \end{bmatrix}^{-1}$$

10.2.3 FINDING THE FREE CONCENTRATIONS [M], [L] AND [H]

The solution of the non-linear equations of mass balance is achieved by using Newton's method (Section 4.7.1). Estimates of [M], [L] and [H] are improved by calculating shifts according to the following equation:

$$\begin{bmatrix} [M]\dfrac{\partial T_M}{\partial [M]} & [L]\dfrac{\partial T_M}{\partial [L]} & [H]\dfrac{\partial T_M}{\partial [H]} \\[2ex] [M]\dfrac{\partial T_L}{\partial [M]} & [L]\dfrac{\partial T_L}{\partial [L]} & [H]\dfrac{\partial T_L}{\partial [H]} \\[2ex] [M]\dfrac{\partial T_H}{\partial [M]} & [L]\dfrac{\partial T_H}{\partial [L]} & [H]\dfrac{\partial T_H}{\partial [H]} \end{bmatrix} \begin{bmatrix} \dfrac{\Delta [M]}{[M]} \\[2ex] \dfrac{\Delta [L]}{[L]} \\[2ex] \dfrac{\Delta [H]}{[H]} \end{bmatrix} = \begin{bmatrix} \Delta T_M \\[2ex] \Delta T_L \\[2ex] \Delta T_H \end{bmatrix}$$

The reason for calculating the relative shifts is that the elements of the coefficient matrix are simple multiples of concentrations, and the matrix is symmetrical since $[L]\partial T_M/\partial [L] = [M]\partial T_L/\partial [M]$, etc. Incidentally, the same coefficients matrix can be used to calculate $\partial [H]/\partial \beta$.

There is a hidden danger—multiple minima (Section 4.4). It can be shown by thermodynamic arguments that the minimum in which all free concentrations are positive is unique, and so it is sufficient to ensure that the concentrations remain positive. This could be done by a general change of variable, but that would slow the calculation down. Shift-cutting can be used as a means of constraining the free concentrations to remain positive, or a temporary change of variable may be used whenever a free concentration threatens to become negative.

The calculation of the free concentrations is iterative, so initial estimates are required. At the first point in a titration curve arbitrary values must be used for all the free concentrations except [H], which can be estimated from the observed electrode potential. Thereafter the values at the previous point in the titration curve may be used.

This calculation must be carried out at each data point, and so it is one of the most time-consuming parts of the program as a whole. Therefore any effort made to get even better initial estimates of the free concentrations is likely to be rewarded by reducing the program's running time. Linear extrapolation has proved to be effective on the second and subsequent main refinement cycles.

10.2.4 DEALING WITH NEGATIVE EQUILIBRIUM CONSTANTS

This problem is perhaps an unexpected one, in that equilibrium constants must, for physical reasons, be positive. However, it arises frequently when incorrect models are being examined, a process that is inevitable when the correct model is not known with any certainty. The problem was first considered by Sillén.

To deal with the problem effectively one must analyse its origins. As just suggested it arises when one is trying to fit data with a wrong model. In this context 'wrong' has a specific meaning. The model may be logically sound, but some of the chemical species postulated may be at such low concentrations as to have a negligible effect on the observed electrode potentials. If that is so, the corresponding equilibrium constant will be impossible to calculate. The fact that it may turn out to be negative is probably due to there being other errors in the model. Systematic errors are the most likely cause. The 'wrong' equilibrium constant becomes negative because that reduces the sum of squares.

If this analysis is correct, the correct treatment is not to constrain the equilibrium constants to be positive, but to reject the model. In this matter we have changed our minds, as MINIQUAD did constrain the equilibrium constants to be positive. However, subsequent experience with SUPERQUAD has shown that the model rejection approach is more satisfactory in that it allows parameters to assume negative values during the refinement and eventually to return to the positive domain.

Allowing the equilibrium constants to become negative has a serious side-effect on the calculation of [M], [L] and [H]. The unique minimum only exists when the equilibrium constants are all positive. When some are negative it may be that all the minima have at least one free concentration negative. In this case the calculation of [M], [L] and [H] will fail. One remedy for this is to increase the Marquardt parameter in the main refinement in order to reduce the parameter shifts.

If a refinement converges with one or more equilibrium constants negative we automatically reject this model. We set the corresponding refinement key to -1 (which means that this parameter will be ignored) and perform the whole refinement again with the new model. In fact we do the same with a positive equilibrium constant if its standard deviation is more than 33% of its value.

10.2.5   THE DANGEROUS PARAMETERS

Currently these are the *initial mass, standard electrode potential* and *burette concentration*. The latter was already used by Sillén in a procedure called 'dirty' to allow for the fact that it is difficult to prepare an alkali solution of exactly known strength. The parameters should only be used when the quantities concerned *cannot* be measured independently to a high enough precision.

Refinement of the standard electrode potential is the easiest to justify. Even when the electrode is calibrated before and after a titration it is difficult to be sure of the precise value to use in the calculation. Errors in electrode calibration can be considered to be systematic errors, and it is true that agreement between two or more titration curves is often improved when the standard electrode potential is refined. However, the refined values should be checked to see that they are within the limits of error of the original determination.

The possibility of refining the initial mass was introduced in response to an experimental exigency. When working with ligands that are difficult to synthesize the reagent used in a titration may not be 100% pure. For example, small peptides isolated from biological sources may be contaminated by salts such as sodium chloride or sodium acetate. Since the quantities available are very small (often a few milligrams) full purification is impossible. The degree of impurity can be determined by refining the initial mass. One common situation is that metal salts are usually contaminated by trace quantities of acid, as they are purified by recrystallization from an acid solution.

An interesting case arose during the investigation of a ligand whose purity was not suspect. On refining the initial mass of this ligand it was found that the effective relative molecular mass was 18 daltons higher than anticipated, suggesting the presence of an unexpected molecule of water of crystallization!

Because of the structure of the algorithm other dangerous parameters could be introduced. For example, the factor RT/nF in the Nernst equation could be replaced by a refinable slope parameter. In fact I have made this modification in a 'private' version of SUPERQUAD which is used for electrode calibration. Other parameters that were included in the first draft of the program were the so-called junction potential correction factors, designed to correct the electrode potentials for non-linear response at low and high pH;

$$E = E^0 + \text{slope} \times \log_e [H] + r \times [H] + s \times [H]^{-1}$$

They were omitted from the final version as it was felt that they were of limited value.

10.2.6 MODEL SELECTION

As stated in the introduction, selecting the best model to represent the experimental data may be the true objective of the equilibrium constant computations. Accordingly a simple model selector has been built into SUPERQUAD. It works by examining a sequence of models, and applying some criteria to decide which is the 'best'. The sequence is generated in two ways. If an equilibrium constant finally becomes negative a new model is generated automatically. Alternatively, one or more new models may be specified at the end of the data input.

A model is defined in terms of the refinement keys of a basic set of equilibrium constants. Each key may be -1, 0 or 1. A vector of these keys therefore defines the model in terms of the basis set. To conserve storage the keys were packed as base 3 integers into integer variables (one 32-bit variable may hold sixteen 2-bit integers). With each new model the defining vector is checked against the models previously refined in order that no model is refined twice. The criteria for accepting a new model are crude: all the calculated equilibrium constants should be positive and have standard deviations of less than 33% of the parameter's value, and the sum of squares should be the lowest up to that stage. A flowchart of the logic is shown in the original publication.

Model selection is the most difficult part of any least-squares analysis, as discussed in Chapters 5 and 6. As yet we have not been able to devise an automatic method for generating candidate models. Sillén had suggested that all possible combinations from a chosen basis set could be examined, but this approach is prohibitively expensive in computer time for all but the simplest of systems, since the number of models rapidly climbs into the hundreds. At the present time scientific intuition is the best guide for selection of candidate models. If we could rationalize the basis of this intuition we would be able to build an automatic model selector.

10.2.7 SYSTEMATIC ERRORS

When we minimize an objective function based on residuals in the observations we might expect that the calculated residuals would show only random variations, corresponding to noise in the data. Experience has shown that this expectation is rarely achieved. An example of a residual plot is also shown in Figure 10.1. The residuals shown are weighted, and so already take into account the greater electrode errors near end-points. Why do the residuals show systematic trends?

We cannot answer this question. If we could, we would have taken appropriate action to eliminate the systematic trends. The best that can be done is to point out possible sources of systematic error and leave it to experiment to find the answer. On the one hand, there are systematic errors in the data. How they come about is a matter for conjecture. The systematic errors are small, to be sure, at a level not very different from the level of random errors. On the other hand, the models used must, at best, be approximate. For example, even when a titration is done in a medium of constant ionic strength, can we be really sure that activity quotients are constant, as implied by the use of equilibrium constants defined as concentration quotients? Are the observational errors truly uncorrelated?

It is a cause for concern, when all the computational problems have apparently been solved and a pukka algorithm has been constructed, that there should be systematic trends in the residuals. How should this be reflected in statements concerning the reliability of the calculated parameters? Are we missing something of fundamental importance? These and other questions may in future be answered. For the moment let me just conclude by saying, 'There are no such things as perfect computer programs, but there are programs that do exactly what their authors say they should do.'

Postscript. SUPERQUAD was originally designed to run on mainframe computers. The data was in fixed format and some users modified the program to use free-format input. Nevertheless, data input was not simple. Output was in the form of tables and line-printer plots. These features were dictated by the fact that the program has to be portable to the maximum possible extent.

With the advent of the newer generation of PCs it is now possible to run the program on a laboratory computer. At the same time the user interface has been improved. My colleague Leslie Pettit has written a SUPERQUAD data editor that has greatly eased the task of data entry. He has also included proper graphical output. This version of the program is distributed in compiled form, something that has become possible now that most people have access to a PC.

In this book I have said nothing about computer hardware and next to nothing about the user interface. However, if a program is to be used as a laboratory tool, the user interface is very important. Ease of data entry and a useful form of results output should be considered as important specifications, to be taken into consideration at the initial stages of program design; it is no good having a good algorithm if it is difficult to use and prone to trivial errors of data input on the user's part.

Appendix 1

Glossary of Statistical Terms

The purpose of this glossary is to give a brief explanation of some of the words and concepts of statistics. For further details the reader is referred to one of the many books dealing with statistics in the physical sciences. The following terms are covered:

Confidence level

Confidence limit

Correlation coefficient

Covariance

Cumulative distribution

Expectation value

Kurtosis

Likelihood (including maximum likelihood)

Mean

Moments

Random variable

Residual

Significance level

Skewness

Standard deviation

Statistic

Variance

Confidence level

The confidence level for a continuous random variable is the probability that a given value will occur in a given interval. In terms of hypothesis testing it gives the probability of a given event occurring if the hypothesis is true. It may be expressed as a fraction in the range 0–1 or as a percentage in the range 0–100%.

Confidence limit

The term confidence limit is applied in place of probability to quantities that are fixed, but bracketed by random variables. This is because a probability can only be attached to a random variable; fixed quantities, such as the mean of a distribution have unit probability. Therefore, when asking questions such as 'what is the probability that the mean will lie between certain values of a random variable?' the term confidence limit is preferred to make it clear that we are not talking about probability in connection with a non-random variable.

Correlation coefficient

The correlation coefficient between two random variables x and y is defined as

$$\rho_{xy} = \frac{(\mathrm{COV})_{xy}}{\sigma_x \sigma_y}$$

Correlation coefficients can take any value between -1 and $+1$, including zero when the variables are uncorrelated.

Covariance
See Appendix 6.

Cumulative distribution function

The probability distribution function of a continuous random variable, $P(y)$, gives the probability that the variable may have a value between y and $y + \mathrm{d}y$. The cumulative distribution function gives the probability that the variable's value will not exceed a given value. It is therefore the integral of the probability distribution function from the lowest value of the random variable up to the given value:

$$F(x) = \int_{-\infty}^{x} P(y) \, \mathrm{d}y$$

For a discrete variable the cumulative probability is simply the sum of probabilities up to the given value of the variable.

Expectation value
See Appendix 6.

Kurtosis

Kurtosis is a property of a probability distribution, derived from the fourth moment of the distribution about the mean. It describes the flatness of the peak in the distribution. The moment coefficient of kurtosis (fourth moment about the mean divided by the square of the variance) of a normal distribution is 3; values less than 3 imply a distribution that is flatter than the normal distribution (platykurtic), while values greater than 3 imply a more pointed distribution (mesokurtic).

Likelihood (including maximum likelihood)

A likelihood function is the product of individual probabilities or of probability distribution functions. The random variable that most concerns us is experimental error. If we assume that experimental error has a normal probability distribution with mean zero and standard deviation σ, the distribution function is

$$P(x_i) = \frac{1}{\sigma(2\pi)^{1/2}} \exp\left(\frac{-x_i^2}{2\sigma^2}\right) \, dx$$

and it follows that the likelihood function is

$$L(x) = \left(\frac{1}{\sigma(2\pi)^{1/2}}\right)^m \exp\left(\frac{-\Sigma x_i^2}{2\sigma^2}\right) \, dx_1 \, dx_2 \, dx_3 \cdots$$

where x_i represents the individual experimental errors, of which there is one for each measurement. To find the maximum likelihood we must make the exponent as small as possible, which means minimizing the sum of the errors squared. In practice the errors are unknowable and we minimize the sum of squared residuals.

Mean
See Appendix 6.

Moments
The nth moment of a random variable x is simply the expectation value of x^n. Of particular interest are the central moments or moments about the mean, i.e. the expectation values of powers of $x - \mu$. The second central moment is the variance. The third and fourth moments are known as skewness and kurtosis, but it is more convenient to use the dimensionless moment coefficients obtained by dividing the moment by the nth power of the standard deviation.

Random variable
A random variable is one whose value cannot be predicted. An experimental variable derived from observations may be the sum of components, some of which are predictable and others of which may be unpredictable. Since such a variable has an unpredictable component it is a random variable.

Residual
A residual is defined as the difference between an observed value and the corresponding calculated value. Residuals are conceptually related to experimental errors, but are quite distinct from them. The method of least squares minimizes the sum of squared residuals. The resulting residuals might be expected to have many of the properties of the experimental errors, such as randomness of sign. An analysis of the residuals is an essential part of any least-squares analysis.

Significance level
The significance level is the opposite of the confidence level; i.e. numerically it is one minus the confidence level (expressed on the scale 0 to 1).

Skewness

Skewness is a property of a probability distribution, derived from the third moment of the distribution about the mean. It describes the extent to which the distribution is symmetrical about the mean. All symmetrical distributions, including the normal distribution, have zero third moments about the mean. The moment coefficient of skewness (third moment about the mean divided by the cube of the standard deviation) is a useful measure of the symmetry of the distribution.

Standard deviation

The standard deviation is defined as the positive square root of the variance.

Statistic

A statistic is any number to which statistical arguments might be applied. Examples include mean, standard deviation, index (FT100, Dow-Jones, cost of living) and moment.

Variance

The variance of a distribution is the expectation value of the square of the difference from the mean; i.e. it is the second moment about the mean. The positive square root of the variance is the standard deviation.

Appendix 2

Some Matrix Algebra

Matrices provide a very compact and useful way of handling linear transformations. This appendix gives definitions of some common terms in matrix algebra, but without proof of any of the relationships. For these and more details the reader should consult the references.

To introduce the concepts let us consider the normal equations for the least-squares determination of three parameters:

$$A_{11}p_1 + A_{12}p_2 + A_{13}p_3 = b_1$$

$$A_{21}p_1 + A_{22}p_2 + A_{23}p_3 = b_2$$

$$A_{31}p_1 + A_{32}p_2 + A_{33}p_3 = b_3$$

This is a linear transformation of the parameters p_1, p_2 and p_3 into b_1, b_2 and b_3. The array of coefficients of p_1, p_2 and p_3 is termed the normal equations matrix and is written as

$$\mathbf{A} = \begin{bmatrix} A_{11} & A_{12} & A_{13} \\ A_{21} & A_{22} & A_{23} \\ A_{31} & A_{32} & A_{33} \end{bmatrix}$$

The general definition of a matrix is an array of numbers with m rows and n columns. This is called an $m \times n$ matrix. The general form of a matrix is *rectangular*, but if $m = n$ the matrix is said to be *square*. The *order* of a square matrix is the number n of rows and columns.

A particular case is the matrix with one column, which is known as a *vector*. For example, the parameters above can be written as a vector:

$$\mathbf{p} = \begin{bmatrix} p_1 \\ p_2 \\ p_3 \end{bmatrix}$$

Notice that the names of matrices and vectors are written in **bold** type.

Matrices and vectors can be *added* and *subtracted*. It is necessary that m and n be the same for both matrices. If that is so, the sum or difference is obtained by combining together corresponding elements:

$$\mathbf{A} \pm \mathbf{B} = \mathbf{C} \text{ implies } C_{ij} = A_{ij} \pm B_{ij} \qquad (i = 1, m, \ j = 1, n)$$

Scalar multiplication, or multiplication by a constant, implies every element of the matrix is multiplied by the constant:

$$\mathbf{C} = c\mathbf{A} \text{ implies } C_{ij} = cA_{ij} \qquad (i = 1, m, \ j = 1, n)$$

Matrix multiplication is defined by a summation of terms:

$$\mathbf{AB} = \mathbf{C} \text{ implies } C_{ij} = \sum_{k=1,n} A_{ik}B_{kj} \qquad (i = 1, m, \ j = 1, n)$$

In this process terms from *row i* of $\mathbf{A}$ are multiplied by terms from *column j* of $\mathbf{B}$, and the products are summed together to give the element of the product in row i and column j. Clearly the number of columns in $\mathbf{A}$ must be the same as the number of rows in $\mathbf{B}$. When that is so the matrices $\mathbf{A}$ and $\mathbf{B}$ are said to be *conformable*. Multiplication of an $m \times n$ matrix with an $n \times m'$ matrix gives a resultant $m \times m'$ matrix; i.e. the result has as many rows as the left-hand matrix and as many columns as the right-hand matrix. Thus the product of a matrix and a vector is a vector. The normal equations can be written as

$$\mathbf{Ap} = \mathbf{b}$$

The product of two square matrices $\mathbf{AB}$ is not necessarily the same as the product $\mathbf{BA}$. If it is the same the matrices are said to *commute*. In the product $(\mathbf{AB})\mathbf{B}$ is said to be left-multiplied by $\mathbf{A}$, and $\mathbf{A}$ is said to be right-multiplied by $\mathbf{B}$.

The *transpose* of a matrix is obtained by interchanging rows and columns. In this book we denote the transpose of $\mathbf{A}$ by $\mathbf{A}^{\mathrm{T}}$:

$$\mathbf{A}^{\mathrm{T}} \text{ implies } (A^{\mathrm{T}})_{ij} = A_{ji} \ (i = 1, m, \ j = 1, n)$$

A matrix that is equal to its own transpose is said to be *symmetric* (and of course symmetric matrices must be square). There are a few special forms of matrix that have individual names. A matrix or vector whose elements are all zero is called a *null* matrix or vector. A matrix of the form

$$\mathbf{A} = \begin{bmatrix} A_{11} & 0 & 0 & \cdots & 0 \\ 0 & A_{22} & 0 & \cdots & 0 \\ 0 & 0 & A_{33} & \cdots & 0 \\ \cdots & \cdots & \cdots & \cdots & 0 \\ 0 & 0 & 0 & 0 & A_{nn} \end{bmatrix}$$

is said to be *diagonal*. If all the elements of a diagonal matrix are equal to unity the matrix is called an *identity* matrix, and given the symbol **E** or **I**. If the matrix has the form

$$\mathbf{A} = \begin{bmatrix} A_{11} & A_{12} & A_{13} & \cdots & A_{1n} \\ 0 & A_{22} & A_{23} & \cdots & A_{2n} \\ 0 & 0 & A_{33} & \cdots & A_{3n} \\ \cdots & \cdots & \cdots & \cdots & \cdots \\ 0 & 0 & 0 & \cdots & A_{nn} \end{bmatrix}$$

it is said to be *upper triangular*. An upper triangular matrix whose diagonal elements A_{jj} are zero is said to be *strict upper triangular*. *Lower triangular* and *strict lower triangular* can be seen as the transposes of the respective upper triangular forms. The *inverse* of a square matrix is defined as follows:

$$\mathbf{AA}^{-1} = \mathbf{A}^{-1}\mathbf{A} = \mathbf{E}$$

Thus the inverse introduces into matrix algebra a process akin to division in scalar algebra. For example, if

$$\mathbf{Ap} = \mathbf{b} \qquad \text{then} \qquad \mathbf{p} = \mathbf{A}^{-1}\mathbf{b}$$

Not all square matrices have an inverse, as we shall see. Matrices whose transpose is equal to the inverse are said to be *orthogonal*. Therefore, if **A** is orthogonal

$$\mathbf{AA}^{\mathrm{T}} = \mathbf{A}^{\mathrm{T}}\mathbf{A} = \mathbf{E}$$

The *determinant* of **A** is denoted by $|\mathbf{A}|$. The determinant can be defined in various ways, none of which make its nature very clear. One rigorous definition is that it is obtained by adding together all the products of n elements of the matrix taken as one from each row and one from each column. Half of the products are also multiplied by -1.[*] For $n = 2$ or 3 the determinant can be written out:

$$\begin{vmatrix} A_{11} & A_{12} \\ A_{21} & A_{22} \end{vmatrix} = A_{11}A_{22} - A_{12}A_{21}$$

$$\begin{vmatrix} A_{11} & A_{12} & A_{13} \\ A_{21} & A_{22} & A_{23} \\ A_{31} & A_{32} & A_{33} \end{vmatrix} = \begin{matrix} A_{11}A_{22}A_{33} + A_{12}A_{23}A_{31} + A_{13}A_{21}A_{32} \\ - A_{11}A_{23}A_{32} - A_{12}A_{21}A_{33} - A_{13}A_{22}A_{31} \end{matrix}$$

---

[*] To find out which ones are positive and which are negative, first arrange the elements in order of the rows from which they are taken.

$$\text{Product} = A_{1i}A_{2j}A_{3k}\cdots$$

The elements are now interchanged in pairs to bring the column indices into ascending numerical order. If the number of interchanges is odd the factor -1 is applied.

The evaluation of determinants of higher order is best performed by one of the specialized techniques, which include Choleski factoring (see below). Determinants have the following properties. They are unchanged if any pair of rows and columns are interchanged or if a multiple of one row (or column) is added to another row (or column). The value of a determinant changes sign if any two rows (or columns) are interchanged. The determinant of an orthogonal matrix is plus or minus unity.

If the determinant of a (square) matrix is zero the matrix is said to be *singular*. The *rank* of a matrix is the size of the largest square matrix, derived by striking out rows and columns, that is not singular. A non-singular square matrix is said to have full *rank* since no rows or columns need to be struck out. Two situations can occur that render a matrix singular: if all elements of a row or column are zero or if all elements of one row are linear combinations of the corresponding elements in other rows or likewise with columns.

The *trace* of a matrix is the sum of its diagonal elements. The trace (or character) of a matrix is invariant to a similarity transformation:

$$\text{Trace}(\mathbf{A}) = \text{trace}(\mathbf{BAB}^{-1})$$

We can now give some general equalities in matrix algebra. In the following $\mathbf{A}$, $\mathbf{B}$, $\mathbf{C}$ are matrices and $\mathbf{a}$ and $\mathbf{b}$ are vectors, c is a scalar:

$$\mathbf{A} + \mathbf{B} = \mathbf{B} + \mathbf{A}$$

$$\mathbf{ABC} = (\mathbf{AB})\mathbf{C} = \mathbf{A}(\mathbf{BC})$$

$$\mathbf{A}c\mathbf{B} = c\mathbf{AB}$$

$$\mathbf{A}(\mathbf{B} + \mathbf{C}) = \mathbf{AB} + \mathbf{AC}$$

$$(\mathbf{ABC}\cdots)^{\mathrm{T}} = \cdots \mathbf{C}^{\mathrm{T}}\mathbf{B}^{\mathrm{T}}\mathbf{A}^{\mathrm{T}}$$

$$(\mathbf{ABC}\cdots)^{-1} = \cdots \mathbf{C}^{-1}\mathbf{B}^{-1}\mathbf{A}^{-1}$$

$$|\mathbf{AB}| = |\mathbf{A}||\mathbf{B}|$$

$$\mathbf{a}^{\mathrm{T}}\mathbf{b} = c$$

$$\mathbf{a}^{\mathrm{T}}\mathbf{Ba} = c \text{ if } \mathbf{B} \text{ is square and order } n$$

Eigenvalues are the solutions to the *characteristic equation* of a square matrix, obtained by expanding the *characteristic determinant:*

$$\begin{vmatrix} A_{11} - \lambda & A_{12} & A_{13} & \cdots & A_{1n} \\ A_{21} & A_{22} - \lambda & A_{23} & \cdots & A_{2n} \\ A_{31} & A_{32} & A_{33} - \lambda & \cdots & A_{3n} \\ \vdots & \cdots & \cdots & \cdots & \vdots \\ A_{n1} & A_{n2} & A_{n3} & \cdots & A_{nn} - \lambda \end{vmatrix}$$

The eigenvalues are the solutions λ_k to the polynomial that results from the expansion:

$$-1^n\lambda^n + -1^{n-1}\lambda^{n-1} \sum_i A_{ii} + \cdots + |\mathbf{A}| = 0$$

From the form of the characteristic equation it can be shown that the product of all the eigenvalues is equal to the value of the determinant of $\mathbf{A}$, and the sum of the eigenvalues is equal to the *trace* of $\mathbf{A}$. From this we may conclude that if a matrix is singular it has one or more zero eigenvalues. If all the eigenvalues are positive the matrix is said to be *positive definite*. It follows that the determinant of a positive definite matrix must be a positive number. The normal equations matrix falls into this category (see Appendix 3). However, it may be that one or more eigenvalues is very small, in which case the matrix may almost be singular. If a matrix is singular it can not have an inverse.

The *eigenvectors*, $\mathbf{h}$, of a matrix are those vectors that obey the relationship

$$\mathbf{Ah} = \lambda\mathbf{h}$$

If the eigenvectors are used as the columns of a matrix $\mathbf{H}$ and $\mathbf{A}$ is symmetric, $\mathbf{H}$ is known as the eigenvector matrix and may be made orthogonal:

$$\mathbf{HAH}^T = \Lambda; \qquad \mathbf{A} = \mathbf{H}^T\Lambda\mathbf{H}$$

where Λ is a diagonal matrix of the eigenvalues of $\mathbf{A}$.

An *idempotent* matrix is one that is equal to its own square:

$$\mathbf{AA} = \mathbf{A}$$

It follows that the eigenvalues of an idempotent matrix are 0 or 1:

$$\mathbf{AA} = \mathbf{H}^T\Lambda\mathbf{HH}^T\Lambda\mathbf{H} = \mathbf{H}^T\Lambda^2\mathbf{H} = \mathbf{H}^T\Lambda\mathbf{H}$$

$$\lambda_i = \lambda_i^2, \qquad i = 1, n$$

One way of calculating the determinant is to perform the Choleski factorization:

$$\mathbf{A} = \mathbf{LDL}^T$$

where $\mathbf{L}$ is lower triangular and has unit diagonal elements, and $\mathbf{D}$ is diagonal. Then

$$|\mathbf{A}| = |\mathbf{LDL}^T| = |\mathbf{L}|\,|\mathbf{D}|\,|\mathbf{L}^T| = |\mathbf{D}|$$

since the determinant of $\mathbf{L}$ is one. Therefore the determinant of $\mathbf{A}$ is given by the product of the elements of $\mathbf{D}$, and if one element of $\mathbf{D}$ is zero or negative the matrix should be regarded as not positive definite. The term positive definite applies strictly only in an algebraic sense. When calculating the determinant numerically the number precision of the computer blurs the definition.

Blocked (partitioned) matrices can be manipulated block by block. For example, if A_{11}, etc., are matrices

$$\left(\begin{array}{c|c} A_{11} & A_{12} \\ \hline A_{21} & A_{22} \end{array}\right)\left(\begin{array}{c|c} B_{11} & B_{12} \\ \hline B_{21} & B_{22} \end{array}\right) = \left(\begin{array}{c|c} A_{11}B_{11} + A_{12}B_{21} & A_{11}B_{12} + A_{12}B_{22} \\ \hline A_{21}B_{11} + A_{22}B_{21} & A_{21}B_{12} + A_{22}B_{22} \end{array}\right)$$

The only requirement is that the blocks be individually conformable. Blocking is particularly useful if one or more blocks are null matrices.

A rectangular $m \times n$ matrix can be partitioned by orthogonal transformations:

$$A = H\left(\begin{array}{c|c} R_{11} & 0 \\ \hline 0 & 0 \end{array}\right)K^T = HRK^T$$

where H is an orthogonal $m \times m$ matrix and K is an orthogonal $n \times n$ matrix. R is an $m \times n$ matrix and R_{11} is a $k \times k$ matrix, where k is the rank of A. R_{11} may be upper triangular, or even diagonal. If it is diagonal it is usually represented by S and is composed of the *singular values*. (The eigenvalues of a square matrix are the squares of the singular values.) In the case that R_{11} is triangular and of rank n the K matrix may be taken as an identity matrix.

By using the orthogonal decomposition above we can define a *pseudo-inverse* for a rectangular matrix:

$$A^+ = K\left(\begin{array}{c|c} R_{11}^{-1} & 0 \\ \hline 0 & 0 \end{array}\right)H^T = KR^+H^T$$

The pseudo-inverse has dimensions $n \times m$ and has the following properties:

$$AA^+A = A$$
$$A^+AA^+ = A^+$$
$$(AA^+)^T = AA^+$$
$$(A^+A)^T = A^+A$$

Appendix 3

Some Properties of the Normal Equations Matrix

This is formed as the product

$$\mathbf{A} = \mathbf{J}^T \mathbf{w}^T \mathbf{w} \mathbf{J}$$

The Jacobian $\mathbf{J}$ has m rows and n columns and is assumed, for the moment, to have rank n; $\mathbf{w}$ is a factor of the weight matrix $\mathbf{W}$. $\mathbf{A}$ is symmetrical; it is also positive definite. This can be seen as follows. Perform a singular value decomposition on $\mathbf{w}\mathbf{J}$:

$$\mathbf{w}\mathbf{J} = \mathbf{H}\left(\frac{\mathbf{S}_{11}}{\mathbf{0}} \left| \frac{\mathbf{0}}{\mathbf{0}}\right.\right)\mathbf{K}^T = \mathbf{HSK}^T$$

$$\mathbf{J}^T \mathbf{w}^T \mathbf{w}\mathbf{J} = \mathbf{KSH}^T \mathbf{HSK}^T = \mathbf{KS}^2\mathbf{K}^T$$

Since eigenvalues are unchanged by orthogonal transformations the eigenvalues of $\mathbf{A}$ are the squares of the singular values of $\mathbf{w}\mathbf{J}$, so all the eigenvalues are positive. If, however, there is a linear dependence between parameters, the rank of $\mathbf{J}$ will be less than n and $\mathbf{A}$ will have $n - k$ zero eigenvalues, one corresponding to each relationship of linear dependence between parameters. Such linear dependences may be adventitious or they may be expressions of equality constraints amongst the parameters. In the latter case the Jacobian does not have full rank.

We will now show that the correlation coefficients ρ_{ij} are restricted to the range

$$-1 < \rho_{ij} < 1$$

The correlation coefficients are obtained from the inverse of the normal equations matrix, $\mathbf{C}(\mathbf{C} = \mathbf{A}^{-1})$:

$$\rho_{ij} = \left(\frac{C_{ij}}{C_{ii}C_{jj}}\right)^{1/2}$$

The condition $-1 < \rho_{ij} < 1$ is equivalent to $1 - \rho_{ij}^2 \geqslant 0$. Since the inverse of a positive definite matrix is also positive definite, and so also has positive diagonal elements, the condition $-1 < \rho_{ij} < 1$ is equivalent to

$$C_{ii}C_{jj} - C_{ij}^2 \geqslant 0, \qquad \text{that is } 1 - \rho_{ij}^2 \geqslant 0$$

These relationships follow from the positive definite nature of $\mathbf{C}$. If $\mathbf{C}$ is positive definite, then $\mathbf{x}^T\mathbf{C}\mathbf{x}$ is a positive scalar for *all* vectors $\mathbf{x}$ (this property can be used as a definition of positive definiteness). Therefore, if $\mathbf{x}$ is a null vector apart from two entries, the quadratic form $\mathbf{x}^T\mathbf{C}\mathbf{x}$ becomes

$$\mathbf{x}^T\mathbf{C}\mathbf{x} = [x_i \ \ x_j] \begin{bmatrix} C_{ii} & C_{ij} \\ C_{ij} & C_{jj} \end{bmatrix} \begin{bmatrix} x_i \\ x_j \end{bmatrix}$$

It follows that all submatrices of the form

$$\begin{bmatrix} C_{ii} & C_{ij} \\ C_{ij} & C_{jj} \end{bmatrix}$$

are positive definite, so $C_{ii}C_{jj} - C_{ij}^2 > 0$.

Appendix 4

Methods for Partial Differentiation

Non-linear least-squares problems require that expressions for the partial derivatives of the model with respect to each parameter can be written down. In this appendix the principal methods for doing this are summarized. Let the model be represented as a function of some parameters, p_k, and independent variables, x:

$$y_i^{\text{calc}} = f(p_k, k = 1, n; x_i)$$

The partial derivatives are defined as follows:

$$\frac{\partial y_i}{\partial p_j} = \lim_{\delta p_j \to 0} \left[\frac{f(\cdots p_j + \delta p_j \cdots; x_i) - f(\cdots p_j \cdots; x_i)}{\delta p_j} \right]$$

In this definition all parameters except p_j are held at constant values. Therefore in practice an expression for a partial derivative with respect to one parameter is evaluated from the model as though only that parameter were variable and all other terms in the model were constants. The ordinary rules of differentiation then apply. In the following a, n are constants and u, v are functions of the variables p, x and z.

Simple expressions
A list of the formulae for the derivatives of simple expressions can be found in reference books such as *The Handbook of Chemistry and Physics* (the 'rubber' book). Some of these are given in Table A4.1.

The chain rule
Let u be a function of x and z, and y be a function of u. Then

$$\frac{\partial y}{\partial x} = \frac{\mathrm{d} y}{\mathrm{d} u} \frac{\partial u}{\partial x}, \qquad \frac{\partial y}{\partial z} = \frac{\mathrm{d} y}{\mathrm{d} u} \frac{\partial u}{\partial z}$$

Table A4.1. Derivatives of simple expressions

Expression	Derivative $\partial/\partial x$
ax^n	anx^{n-1}
$\sin(x)$	$\cos(x)$
$\cos(x)$	$-\sin(x)$
$\tan(x)$	$1/\cos^2(x)$
$\log_e(x)$	$1/x$
e^{ax}	$a\,e^{ax}$
$u^n v^n$	$u^{n-1}v^{n-1}\left(nv\,\dfrac{\partial u}{\partial x} + mu\,\dfrac{\partial v}{\partial x}\right)$
$\dfrac{u^n}{v^n}$	$\dfrac{u^{n-1}}{v^{n+1}}\left(nv\,\dfrac{\partial u}{\partial x} - mu\,\dfrac{\partial v}{\partial x}\right)$

For example,

$$y = e^{(xz)^2}; \qquad u = (xz)^2; \qquad y = e^u$$

$$\frac{dy}{du} = e^u; \qquad \frac{\partial u}{\partial x} = 2(xz)z; \qquad \frac{\partial u}{\partial z} = 2(xz)x$$

$$\frac{\partial y}{\partial x} = e^{(xz)^2}(2xz^2); \qquad \frac{\partial y}{\partial z} = e^{(xz)^2}(2x^2z)$$

This example could also be done with the substitutions $v = xz$, $u = v^2$:

$$\frac{\partial y}{\partial x} = \frac{dy}{du}\frac{du}{dv}\frac{\partial v}{\partial x} = e^u(2v)z$$

This shows how repeated use of the chain rule can break down complicated expressions into simple ones.

Implicit differentiation

Suppose that we have two equations

$$a = x + pxz$$
$$b = z + pxz$$

and we wish to find an expression for $\partial x/\partial p$. In this case x is said to be an implicit function of p. First we must expand the equations into a first-order Taylor series:

$$\delta a = \frac{\partial a}{\partial p}\,\delta p + \frac{\partial a}{\partial x}\,\delta x + \frac{\partial a}{\partial z}\,\delta z$$

$$\delta b = \frac{\partial b}{\partial p}\,\delta p + \frac{\partial b}{\partial x}\,\delta x + \frac{\partial b}{\partial z}\,\delta z$$

Then, we divide throughout by δp and take the limit $\delta p \to 0$:

$$\frac{\partial a}{\partial p} = \frac{\partial a}{\partial x}\frac{\partial x}{\partial p} + \frac{\partial a}{\partial z}\frac{\partial z}{\partial p}$$

$$\frac{\partial b}{\partial p} = \frac{\partial b}{\partial x}\frac{\partial x}{\partial p} + \frac{\partial b}{\partial z}\frac{\partial z}{\partial p}$$

Thus, the required derivative $\partial x/\partial p$, along with $\partial z/\partial p$, is seen to be a solution of the two simultaneous equations, whose terms consist of the partial derivatives of a and b with respect to p, x and z. The latter are obtained from the defining equations in which a and b are explicit functions of p, x and z.

In this case we may write the solution down, but had there been three or more equations it would become very tedious to do so. However, the *value* of the derivative can easily be found by a computer program that solves the simultaneous equations, so in this case we have to replace an expression from which the value of the derivative may be found by a recipe specifying the evaluation of a number of expressions.

Differentiation of integrals
The derivative of the definite integral

$$I(z) = \int_a^b f(x, z)\,\mathrm{d}x$$

is given by

$$\frac{\mathrm{d}I(z)}{\mathrm{d}z} = f(b, z)\frac{\mathrm{d}b}{\mathrm{d}z} - f(a, z)\frac{\mathrm{d}a}{\mathrm{d}z} + \int_a^b \frac{\partial f(x, z)}{\partial z}\,\mathrm{d}x$$

If the integration limits a and b do not depend on z this becomes

$$\frac{\mathrm{d}I(z)}{\mathrm{d}z} = \int_a^b \frac{\partial f(x, z)}{\partial z}\,\mathrm{d}x$$

In some cases these formulae can be used when one or both integration limits are infinite, but it is not always so. The interested reader is referred to books on advanced calculus.

In summary, it should be noted that if a relationship between two quantities can be written down as an equation the partial derivative *can* be found, with one proviso. This is that all the quantities involved must be continuous and have continuous derivatives at the point where the evaluation is to be performed or that integrals should exist over the integration range.

Appendix 5

Proof of the Expected Value of the Sum of Squared Residuals

In Section 6.4.1 it was required to show that the expected value of the sum of squared weighted residuals given by

$$S = (\mathbf{y}^{obs})^T \mathbf{W} \mathbf{y}^{obs} - \mathbf{p}^T (\mathbf{J}^T \mathbf{W} \mathbf{J}) \mathbf{p}$$

is $m - n$. The proof relies heavily on the concept of expectation values (see Appendix 6). For any linear model the vector of observations can be written as

$$\mathbf{y}^{obs} = \mathbf{J}\mathbf{p} + \mathbf{e}$$

where $\mathbf{e}$ is a vector of errors, which are assumed to have expectation values of zero, and $\mathbf{p}$ is a vector of parameters. The expectation values for the observations, $\mathbf{y}^{exp}$, are related to the expectation values for the parameters, $\mathbf{p}^{exp}$, by

$$\mathbf{y}^{exp} = \mathbf{J}\mathbf{p}^{exp}$$

We shall first show that

$$S = (\mathbf{y}^{obs} - \mathbf{y}^{exp})^T \mathbf{W}(\mathbf{y}^{obs} - \mathbf{y}^{exp}) - (\mathbf{p} - \mathbf{p}^{exp})^T \mathbf{J}^T \mathbf{W} \mathbf{J}(\mathbf{p} - \mathbf{p}^{exp})$$

Using the fact that $\mathbf{J}^T \mathbf{W} \mathbf{J}\mathbf{p} = \mathbf{J}^T \mathbf{W} \mathbf{y}^{obs}$ and $\mathbf{J}^T \mathbf{W} \mathbf{J}\mathbf{p}^{exp} = \mathbf{J}^T \mathbf{W} \mathbf{y}^{exp}$,

$$(\mathbf{y}^{obs})^T \mathbf{W} \mathbf{y}^{exp} = (\mathbf{y}^{obs})^T \mathbf{W} \mathbf{J}\mathbf{p}^{exp} = \mathbf{p}^T (\mathbf{J}^T \mathbf{W} \mathbf{J})\mathbf{p}^{exp}$$

$$(\mathbf{y}^{exp})^T \mathbf{W} \mathbf{y}^{obs} = (\mathbf{y}^{exp})^T \mathbf{W} \mathbf{J}\mathbf{p} = (\mathbf{p}^{exp})^T (\mathbf{J}^T \mathbf{W} \mathbf{J})\mathbf{p}$$

$$(\mathbf{y}^{exp})^T \mathbf{W} \mathbf{y}^{exp} = (\mathbf{y}^{exp})^T \mathbf{W} \mathbf{J}\mathbf{p}^{exp} = (\mathbf{p}^{exp})^T (\mathbf{J}^T \mathbf{W} \mathbf{J})\mathbf{p}^{exp}$$

Therefore, six of the terms in S^2 cancel each other in pairs, leaving S equal to the first expression given above.

The two terms are now in the form $(\mathbf{x} - \bar{\mathbf{x}})^{\mathrm{T}}\mathbf{M}^{-1}(\mathbf{x} - \bar{\mathbf{x}})$, where $\mathbf{x}$ is a vector of random variables, $\bar{\mathbf{x}}$ is a vector of their expectation values and $\mathbf{M}$ is the associated variance–covariance matrix. This expression can be rearranged:

$$\sum_i \sum_j (x - \bar{x})_i M_{ij}^{-1} (x - \bar{x})_j = \sum \sum (x - \bar{x})_i (x - \bar{x})_j M_{ij}^{-1}$$

$$= \sum_i [(\mathbf{x} - \bar{\mathbf{x}})(\mathbf{x} - \bar{\mathbf{x}})^{\mathrm{T}}\mathbf{M}^{-1}]_{ii}$$

By definition $(\mathbf{x} - \bar{\mathbf{x}})(\mathbf{x} - \bar{\mathbf{x}})^{\mathrm{T}}$ is the expectation value of the variance–covariance matrix $\mathbf{M}$ associated with the random variables, so $(\mathbf{x} - \bar{\mathbf{x}})(\mathbf{x} - \bar{\mathbf{x}})^{\mathrm{T}}\mathbf{M}^{-1}$ is a unit matrix. The expectation value for $(\mathbf{x} - \bar{\mathbf{x}})^{\mathrm{T}}\mathbf{M}^{-1}(\mathbf{x} - \bar{\mathbf{x}})$ is therefore equal to the sum of the diagonal elements of the unit matrix, which is equal to its order, or rank.

The expectation value of $(\mathbf{y}^{\mathrm{obs}} - \mathbf{y}^{\mathrm{exp}})^{\mathrm{T}}\mathbf{W}(\mathbf{y}^{\mathrm{obs}} - \mathbf{y}^{\mathrm{exp}})$ is equal to the number of observations, m, and the expectation value of $(\mathbf{p} - \mathbf{p}^{\mathrm{exp}})^{\mathrm{T}}\mathbf{J}^{\mathrm{T}}\mathbf{W}\mathbf{J}(\mathbf{p} - \mathbf{p}^{\mathrm{exp}})$ is equal to the number of parameters, n. Therefore the expectation value of S is $m - n$.

Appendix 6

Expectation Values

The expectation value of a random variable x is defined as the sum over all possible values of x of the product of x and its probability:

$$E\{x\} = \Sigma \, xp(x)$$

By definition this expectation value is called the *mean* of x, $\mu(x)$. It is assumed that the probability is normalized so that the sum of all probabilities is unity. This being the case the mean can be seen as a weighted average, the weights being the probabilities. When x is a continuous variable the probability $p(x)$ is replaced by $p(x) \, dx$, and the summation by a definite integral between the limits of existence of x, e.g.

$$E\{x\} = \int_{-\infty}^{\infty} xp(x) \, dx$$

The expectation value of a function of x, $f(x)$, is defined in a similar way:

$$E\{f(x)\} = \Sigma \, f(x)p(x)$$

Variance is defined as the expectation value of the function $[x - \mu(x)]^2$:

$$\sigma^2(x) = E\{[x - \mu(x)]^2\} = \Sigma \, [x - \mu(x)]^2 p(x)$$
$$= E\{x^2\} - \mu\{x\}^2$$

Covariance has to be considered when two random variables are not statistically independent:

$$COV(x, y) = E\{[x - \mu(x)] \, [y - \mu(y)]\}$$
$$= E\{xy\} - \mu\{x\}\mu\{y\}$$

This definition embraces that of the variance since $COV(x, x) = \sigma^2(x)$.

An important property of expectation values is the linear property. If two or more random variables are combined linearly the expectation values also combine linearly:

$$z = ax + by$$

$$E\{z\} = aE\{x\} + bE\{y\}$$

This follows directly from the definition of expectation values. It can be applied to the general linear transformation

$$z_k = \Sigma \, T_{ik} x_i$$

where T_{ik} is a set of transformation coefficients and x_i is a set of random variables. The means of the new variables z_k are given by

$$\begin{aligned}
\mu(z_k) &= E\{z_k\} \\
&= E\{\Sigma \, T_{ik} x_i\} \\
&= \Sigma \, E\{T_{ik} x_k\} \\
&= \Sigma \, T_{ik}\mu(x_i)
\end{aligned}$$

and the covariances are obtained in a similar fashion:

$$\begin{aligned}
\mathrm{COV}(z_i, z_j) &= E\{[z_i - \mu(z_i)][z_j - \mu(z_j)]\} \\
&= E\left\{\left[\sum_h T_{ih} x_h - \sum_h T_{ih}\mu(x_h)\right]\left[\sum_k T_{jk} x_k - \sum_k T_{jk}\mu(x_k)\right]\right\} \\
&= E\left\{\left[\sum_h T_{ih}(x_h - \mu(x_h))\right]\left[\sum_k T_{jk}(x_k - \mu(x_k))\right]\right\} \\
&= \sum_h \sum_k T_{ih} T_{jk} E\{[x_h - \mu(x_h)][x_k - \mu(x_k)]\} \\
&= \Sigma\Sigma \, T_{ih} T_{jk} \, \mathrm{COV}(x_h, x_k)
\end{aligned}$$

This expression takes a simple form in matrix notation:

$$\mathbf{COV(z) = T \; COV(x)T}^{\mathrm{T}}$$

It is the basis for all error-propagation formulae and is particularly useful as *it is independent of the form of the probability distribution function.*

Applications
Expectation values may be used to calculate the values of the mean and variance of any given probability distribution function. For example, the probability distribution function for a uniform random variable, $-a \leqslant x \leqslant a$, is $dx/2a$. Hence

$$\mu(x) = \int_{-a}^{a} \frac{x}{2a} \, dx = 0$$

$$\sigma^2(x) = \int_{-a}^{a} \frac{x^2}{2a} \, dx = \frac{a^2}{3}$$

When random variables are combined linearly the mean and variance of the combination can be calculated without a knowledge of the original distribution functions. A particular case of general importance is that in which the original variables are uncorrelated, that is $\text{COV}(x_i, x_j) = 0$ when $i \neq j$. Then

$$z_k = \Sigma \, T_{ik} x_i$$
$$\mu(z_k) = \Sigma \, T_{ik}\mu(x_i)$$
$$\sigma^2(z_k) = \Sigma \, T_{ki}^2 \sigma^2(x_i)$$

If a set of random variables belonging to a multivariate normal distribution is combined together linearly, then statistical theory shows the resultant combinations also belong to a multivariate normal distribution. The means and covariances of this distribution would be obtained by using the formulae above.

Sample statistics
When we make repeated independent measurements of a variable subject to random experimental error we may derive estimates of the population mean and variance of that variable. Suppose we have made m measurements of a variable y. Denoting the individual values as y_i $(i = 1, m)$, the sample mean as $\bar{y}$ and the population mean as $\mu(y)$,

$$\bar{y} = (1/m)\Sigma \, y_i$$
$$E\{\bar{y}\} = (1/m)\Sigma \, E\{y\}$$
$$= (1/m)\Sigma \, \mu(y)$$
$$= \mu(y)$$

The sample mean is therefore an estimator of the population mean; it is said to be *unbiased* because its expectation value is the same as the population value. If we define the sample variance as

$$s^2 = \frac{\Sigma \, (y_i - \bar{y})^2}{m - 1}$$

it is an unbiased estimator of the population variance σ^2. The proof is as follows. From the definition of variance given above,

$$E\{y^2\} = \sigma^2(y) + \mu^2(y)$$
$$E\{\bar{y}^2\} = \sigma^2(\bar{y}) + \mu^2(\bar{y})$$
$$\sigma^2(\bar{y}) = \sigma^2(\Sigma \, y/m) = \Sigma \, (1^2/m^2)\sigma^2(y) = \sigma^2(y)/m$$
$$E\{\bar{y}^2\} = \sigma^2(y)/m + \mu^2(y)$$

Also, from the definition of covariance,

$$E\{y\bar{y}\} = COV(y\bar{y}) + \mu(y)\mu(\bar{y})$$
$$= \sigma^2(y)/m + \mu^2(y)$$

Therefore,

$$E\{s^2\} = \frac{\Sigma E(y_i)^2 - 2E\{y_i\bar{y}\} + E\{\bar{y}^2\}}{m - 1}$$
$$= \sigma^2(y)$$

In a similar way, the sample covariance between two correlated variables x and y is

$$COV(x, y) = \frac{\Sigma (x_i - \bar{x})(y_i - \bar{y})}{m - 1}$$

Appendix 7

Some Properties of Convolution Functions

In this appendix we outline some of the properties of convolution functions derived by piecewise fitting of a polynomial to a subset of the data, as in Section 7.4:

1. The sum of coefficients of a smoothing function is unity. The sum of coefficients of a derivative function is zero.

   Proof. The convolution coefficients are given by the formula $(\mathbf{J}^T\mathbf{J})^{-1}\mathbf{J}^T$, and the first column of $\mathbf{J}$ is a column of ones. Therefore the sums of convolution coefficients are given by the first column of $(\mathbf{J}^T\mathbf{J})^{-1}\mathbf{J}^T\mathbf{J}$, which is the first column of an identity matrix. It follows that the first sum (that of the smoothing coefficients) is unity and other sums are all zero.

2. Smoothing of a function leaves the area under that function unchanged.

   Proof. Let the smoothing function be discrete and have coefficients f_k; $k = -p \cdots p$, $p = (m-1)/2$. Let the function being smoothed be $g(x)$. The convoluted function, $c(x)$, is then given by

$$c(x) = \sum_{k=-p,p} f_k g(x-k)$$

The area under this function is simply

$$A = \int_{-\infty}^{\infty} c(x)\, dx$$

$$= \int \left[\sum f_k g(x-k) \right] dx$$

$$= \sum f_k \left[\int g(x-k)\, dx \right]$$

$$= \sum f_k \left[\int g(x) \, dx \right]$$

$$= \left[\int g(x) \, dx \right] \sum f_k$$

By property 1 the sum of smoothing coefficients is unity. Therefore the area under the convoluted function is the same as the area under the original function.

A corollary of this is pertinent to the distortion of functions such as Lorentzian and Gaussian bands. If smoothing reduces the height of a band it must, in order to conserve area, increase the width of the band. This is the significance of the negative residuals in Figure 7.3.

The area under a function will be conserved under convolution by a continuous convolution function as long as the area under the latter is unity. Since the sum of derivative convolution coefficients is zero, when a derivative is calculated by convolution the total area under the derivative must be zero. It follows that the area under those parts of the derivative that are positive must be equal to the area over those parts of the derivative that are negative.

3. Convolution of symmetric functions by even-derivative functions conserves the positions of centres of symmetry.

Proof. Let the function, $g(x)$, have a centre of symmetry at $x = x_0$; this function must have a maximum or minimum value at that point. Then

$$\frac{dg(x_0)}{dx} = 0$$

$$\frac{dg(x_0 - k)}{dx} = \frac{-dg(x_0 + k)}{dx}$$

The convolution functions for even derivatives, f_k, are symmetric: $f_k = f_{-k}$. Since

$$c(x) = \sum_{k=-p,p} f_k g(x - k)$$

$$\frac{dc(x)}{dx} = \sum f_k \frac{dg(x - k)}{dx}$$

Since

$$f_k \frac{dg(x_0 - k)}{dx} = -f_{-k} \frac{dg(x_0 + k)}{dx}$$

$$f_0 \frac{dg(x_0)}{dx} = 0$$

the derivative of the convoluted function is zero at $x = x_0$, so that it has an extremum at the same place as the original function. A geometric illustration of this property is shown in Figure 9.1; the area of maximum overlap occurs when the two centres of symmetry coincide. In consequence of this property the maxima of symmetric peaks are not shifted by smoothing, the second derivative has a minimum where the band has a maximum, etc.

Some spectrophotometer manufacturers have utilized this property (which applies to all symmetric convolution functions) to reduce the apparent noise level in spectroscopic curves. The advantage over RC (resistance–capacitance) damping is that the latter is equivalent to convolution with an unsymmetrical function (a one-sided exponential decay), and so causes maxima to be displaced, whereas convolution with symmetrical filter functions does not. As both RC damping and polynomial smoothing are forms of convolution, the data errors will be correlated in both cases (Section 7.4.6). Both processes work well if the noise varies much more rapidly than the signal; if that is not so, some distortion of the signal is inevitable. Correlation and/or distortion should be taken into account when fitting such data by the method of least squares.

References and Additional Reading

Acton F. S. (1959) *Analysis of Straight-Line Data*, Wiley, New York.

Alder, H. L. and E. B. Roessler (1968) *Introduction to Probability and Statistics*, 4th edition, W. H. Freeman, San Francisco.

Baker, C., I. P. Cockerill, J. E. Kelsey and W. F. Maddams (1978) *Spectrochim. Acta*, **34A**, 673.

Baker, C., P. S. Johnson and W. F. Maddams (1978) *Spectrochim. Acta*, **34A**, 683.

Baker, C., W. F. Maddams, J. G . Grasselli and M. A. S. Hazle (1978) *Spectrochim. Acta*, **34A**, 683.

Barker, B. E. and M. F. Fox (1981) *Chem. Soc. Rev.*, **9**, 143.

Barnett, V. (1983) Principles and methods for handling outliers in data sets, in T. Wright (editor), *Statistical Methods and Improvement of Data Quality*, Academic Press, London.

Barnett, V. and T. Lewis (1978) *Outliers in Statistical Data*, Wiley, Chichester.

Bartell, L. S., D. J. Romenesko and T. C. Wong (1975) *Molecular Structure by Diffraction Methods*, Volume 3, p. 72.

Bartels, R. H. and J. J. Jeziorawski (1985) *Assoc. Comp. Machinery Trans. Math. Software*, **11**, 201.

Bauer, E. L. (1971) *A Statistical Manual for Chemists*, 2nd edition, Academic Press, New York.

Bevington, P. R. (1969) *Data Reduction and Error Analysis for the Physical Sciences*, McGraw-Hill, New York.

Bicking, C. A. (1978) Principles and methods of sampling, in I. M. Kolthoff and P. J. Elving (editors), *Treatise on Analytical Chemistry*, 2nd edition, Part 1, Volume 1, pp. 299–357, Wiley, New York.

Blackburn, J. A. (editor) (1970) *Spectral Analysis: Methods and Techniques*, Marcel Dekker, New York.

Blass, W. E. and G. W. Halsey (1981) *Deconvolution of Absorption Spectra*, Academic Press, New York.

Borowiak, D. S. (1989) *Model Discrimination for Non-linear Regression Models*, Marcel Dekker, New York.

Bower, D. I., R. S. Jackson and W. F. Maddams (1990) *J. Polymer Phys.*, **28**, 837.

Box, M. J., D. Davies and W. H. Swann (1969) *Non-Linear Optimization Techniques*, Oliver and Boyd, Edinburgh.

Braibanti, A., G. Ostacoli, P. Paoletti, L. D. Pettit and S. Sammartano (1987) *Pure Appl. Chem.*, **59**, 1721.

Chambers, J. M. (1977) *Computational Methods for Data Analysis*, Wiley, New York.

Champeney, D. C. (1973) *Fourier Transforms and Their Physical Applications*, Academic Press, New York.

Christopher, R. E. and P. Gans (1975) *J. Chem. Soc. Dalton Trans.*, 153.

Clifford, A. A. (1973) *Multivariate Error Analysis*, Applied Science Publishers, London.

Cooper, B. E. (1969) *Statistics for Experimentalists*, Pergamon Press, Oxford.

Cooper, M. J. (1977) *Physics Bull.*, 463.

Cumming, G. L., J. S. Rollet, F. J. C. Rossotti and R. J. Whewell (1972) *J. Chem. Soc. Dalton Trans.*, 2653.

Currie, L. A. (1978) Sources of error and the approach to accuracy, in I. M. Kolthoff and P. J. Elving (editors), *Treatise on Analytical Chemistry*, 2nd edition, Part 1, Volume 1, pp. 95–233, Wiley, New York.

Daniels, F. and E. H. Johnston (1921) *J. Am. Chem. Soc.*, **43**, 53.

de Boor, C. (1978) *A Practical Guide to Splines*, Springer-Verlag, New York.

Deming, W. E. (1943) *Statistical Adjustment of Data*, Wiley, New York.

Dyrssen, D., N. Ingri and L. G. Sillén (1961) *Acta Chem. Scand.*, **15**, 694.

Enke, C. G. and T. A. Niemann (1976) *Analytical Chem.*, **48**, 705A.

Fletcher, R. (1971) UKAEA Report AERE-R 6799, H.M. Stationery Office, London.

Gans, P. (1975) *J. Chem. Soc. Dalton. Trans.*, 153.

Gans, P. and J. B. Gill (1980) *Anal. Chem.*, **52**, 351.

Gans, P. and J. B. Gill (1983) *Appl. Spectrosc.*, 37, 515

Gans, P. and J. B. Gill (1984) *Appl. Spectrosc.*, **38**, 370.

Gans, P., A. Sabatini and A. Vacca (1976) *Inorg. Chim. Acta*, **18**, 239.

Gans, P., A. Sabatini and A. Vacca (1985) *J. Chem. Soc. Dalton Trans.*, 1195.

Gans, P. and A. Vacca (1974) *Talanta*, **21**, 45.

Girard, A. (1958) *Rev. Opt.*, **37**, 225, 397.

Golub, G. H. and C. F. van Loan (1980) *Soc. Ind. Appl. Math. Numer. Anal.*, **17**, 883.

Guest, P. G. (1961) *Numerical Methods of Curve Fitting*, Cambridge University Press.

Hamilton, W. C. (1964) *Statistics in Physical Science*, The Ronald Press, New York.

Hogg, R. V. and A. T. Craig (1970) *Introduction to Mathematical Statistics*, 3rd edition, The Macmillan Company, New York.

Ingri, N. and L. G. Sillén (1964) *Arkiv Kemi*, **23**, 97.

Johnson, L. J. and M. D. Harmony (1973) *Anal. Chem.*, **45**, 1494.

Kolthoff, I. M. and P. J. Elving (editors) (1978) *Treatise on Analytical Chemistry*, 2nd edition, Part 1, Volume 1, Wiley, New York.

Krogh, F. T. (1974) *Comm. Assoc. Comp. Mach.*, **3**, 167.

Lawson, C. L. and R. J. Hanson (1974) *Solving Least Squares Problems*, Prentice-Hall, Englewood Cliffs, New Jersey.

Leggett, D. J. (editor) (1985) *Computational Methods for the Determination of Formation Constants*, Plenum Press, New York.

Levenberg, K. (1944) *Quart. Appl. Math.*, **2**, 164.

Lill, S. A. (1970) *Computer J.*, **13**, 111.

Maddams, W. F. (1980) The scope and limitations of curve fitting, *Appl. Spectrosc.*, **34**, 245.

Madden, H. H. (1978) *Anal. Chem.*, **50**, 1383.

Mandel, J. (1964) *The Statistical Analysis of Experimental Data*, Interscience, New York.

Mandel, J. (1978) Accuracy and precision: evaluation and interpretation of analytical results, in I. M. Kolthoff and P. J. Elving (editors), *Treatise on Analytical Chemistry*, 2nd edition, Part 1, Volume 1, pp. 243–297, Wiley, New York.

Malinowski, E. R. and D. G. Howery (1989) *Factor Analysis in Chemistry*, Robert E. Krieger Publishing Company, Malabar, Florida.

Mardia, K. V., J. T. Kent and J. M. Bibby (1979) *Multivariate Analysis*, Academic Press, London.

Mardia, K. V. and P. J. Zemroch (1978) *Tables of the F- and Related Distributions with Algorithms*, Academic Press, London.

Margenau, H. and G. M. Murpy (1943) *The Mathematics of Physics and Chemistry*, Van Nostrand, Princeton, New Jersey.

Marquardt, D. W. (1963) *J. Soc. Ind. Appl. Math.*, **11**, 431.

Morrison, D. D. (1960) *JPL Seminar Proceedings*.

Nagypal, I., I. Paka and L. Zekany (1987) *Talanta*, **25**, 549.

O'Haver, T. C. and T. Begley (1981) *Anal. Chem.*, **53**, 1876.

O'Neill, R. (1971) *Appl. Statistics*, **20**, 338.

Pentz, M. and M. Shott (1988) *Handling Experimental Data*, Open University Press, Milton Keynes.

Perrin, D. D. and I. G. Sayce (1967) *Talanta*, **14**, 82.

Perrin, D. D. and H. Stunzi (1985) Chapter 4 in D. J. Leggett (editor), *Computational Methods in the Determination of Formation Constants*, Plenum Press, New York.

Powell, M. J. D. (1964) *Computer J.*, **7**, 155.

Press, W. H., B. P. Flannery, S. A. Teukolsky and W. T. Vetterling (1986) *Numerical Recipes*, Cambridge University Press, Cambridge.

Proctor, A. and P. M. A. Sherwood (1980) *Anal. Chem.*, **52**, 2315.

Sabatini, A., A. Vacca and P. Gans (1974) *Talanta*, **21**, 53.

Sabatini, A. and A. Vacca (1972) *J. Chem. Soc. Dalton Trans.*, 1693.

Sadler, D. R. (1975) *Numerical Methods for Nonlinear Regression*, University of Queensland Press.

Savitzsky, A. and M. J. E. Golay (1964) *Anal. Chem.*, **36**, 1627; for some corrections see J. Steinier, Y. Termonia and J. Deltour (1972) *Anal. Chem.*, **44**, 1906.

Sayce, I. G. (1968) *Talanta*, **15**, 1397.

Sayce, I. G. (1971) *Talanta*, **18**, 653.

Sayce, I. G. and V. S. Sharma (1972) *Talanta*, **19**, 831.

Sillén, L. G. (1962) *Acta Chem. Scand.*, **16**, 159; the last paper in this series is P. Brauner, L. G. Sillén and R. Whiteker (1969) *Arkiv Kemi*, **31**, 365.

Spiegel, M. R. (1972) *Theory and Problems of Statistics*, Schaum's Outline Series, McGraw-Hill, New York.

Stability Constants, IUPAC Publications, Pergamon, Oxford.

Stephenson, G. (1961) *Mathematical Methods for Science Students*, Longman's, London.

Stewart, G. W. (1967) *J. Assoc. Comp. Mach.*, **14**, 72.

Sundius, T. (1973) *J. Raman Spectrosc.*, **1**, 471.

Tobias, R. S. and M. Yasuda (1963) *Inorg. Chem.*, **2**, 1307.

Topping, J. (1972) *Errors of Observation and Their Treatment*, Chapman and Hall, London.

Vanderginste, B. G. M. and L. de Galan (1975) *Anal. Chem.*, **47**, 2124.

Willson, P. D. and T. H. Edwards (1976) *Appl. Spectrosc. Rev.*, **12**, 1.

Wynne, C. G. (1959) *Proc. Phys. Soc. Lond.*, **73**, 777.

Index